DISCARD

Leading the Mathematical Sciences Department

A Resource for Chairs

Copyright ©2005 by
The Mathematical Association of America (Incorporated)

Library of Congress Catalog Card Number 2004113542

ISBN 0-88385-174-1

Printed in the United States of America

Current printing (last digit):
10 9 8 7 6 5 4 3 2 1

Leading the Mathematical Sciences Department

A Resource for Chairs

Edited by

Tina H. Straley
Executive Director
Mathematical Association of America
Professor Emerita, Kennesaw State University

Marcia P. Sward
Executive Director Emerita
Mathematical Association of America

Jon W. Scott
Professor, Montgomery College

Published and Distributed by
The Mathematical Association of America

The MAA Notes Series, started in 1982, addresses a broad range of topics and themes of interest to all who are involved with undergraduate mathematics. The volumes in this series are readable, informative, and useful, and help the mathematical community keep up with developments of importance to mathematics.

Committee on Publications
Roger Nelsen, *Chair*

Notes Editorial Board
Sr. Barbara E. Reynolds, *Editor*
Paul E. Fishback, *Associate Editor*

Jack Bookman   Annalisa Crannell   Daniel John Curtin
Rosalie Dance   William E. Fenton   Richard D. Järvinen
Sharon C. Ross

MAA Notes

11. Keys to Improved Instruction by Teaching Assistants and Part-Time Instructors, *Committee on Teaching Assistants and Part-Time Instructors, Bettye Anne Case,* Editor.
13. Reshaping College Mathematics, *Committee on the Undergraduate Program in Mathematics, Lynn A. Steen,* Editor.
14. Mathematical Writing, by *Donald E. Knuth, Tracy Larrabee, and Paul M. Roberts.*
16. Using Writing to Teach Mathematics, *Andrew Sterrett,* Editor.
17. Priming the Calculus Pump: Innovations and Resources, *Committee on Calculus Reform and the First Two Years,* a subcommittee of the Committee on the Undergraduate Program in Mathematics, *Thomas W. Tucker,* Editor.
18. Models for Undergraduate Research in Mathematics, *Lester Senechal,* Editor.
19. Visualization in Teaching and Learning Mathematics, *Committee on Computers in Mathematics Education, Steve Cunningham and Walter S. Zimmermann,* Editors.
20. The Laboratory Approach to Teaching Calculus, *L. Carl Leinbach et al.,* Editors.
21. Perspectives on Contemporary Statistics, *David C. Hoaglin and David S. Moore,* Editors.
22. Heeding the Call for Change: Suggestions for Curricular Action, *Lynn A. Steen,* Editor.
24. Symbolic Computation in Undergraduate Mathematics Education, *Zaven A. Karian,* Editor.
25. The Concept of Function: Aspects of Epistemology and Pedagogy, *Guershon Harel and Ed Dubinsky,* Editors.
26. Statistics for the Twenty-First Century, *Florence and Sheldon Gordon,* Editors.
27. Resources for Calculus Collection, Volume 1: Learning by Discovery: A Lab Manual for Calculus, *Anita E. Solow,* Editor.
28. Resources for Calculus Collection, Volume 2: Calculus Problems for a New Century, *Robert Fraga,* Editor.
29. Resources for Calculus Collection, Volume 3: Applications of Calculus, *Philip Straffin,* Editor.
30. Resources for Calculus Collection, Volume 4: Problems for Student Investigation, *Michael B. Jackson and John R. Ramsay,* Editors.
31. Resources for Calculus Collection, Volume 5: Readings for Calculus, *Underwood Dudley,* Editor.
32. Essays in Humanistic Mathematics, *Alvin White,* Editor.
33. Research Issues in Undergraduate Mathematics Learning: Preliminary Analyses and Results, *James J. Kaput and Ed Dubinsky,* Editors.
34. In Eves' Circles, *Joby Milo Anthony,* Editor.
35. You're the Professor, What Next? Ideas and Resources for Preparing College Teachers, *The Committee on Preparation for College Teaching, Bettye Anne Case,* Editor.
36. Preparing for a New Calculus: Conference Proceedings, *Anita E. Solow,* Editor.
37. A Practical Guide to Cooperative Learning in Collegiate Mathematics, *Nancy L. Hagelgans, Barbara E. Reynolds, SDS, Keith Schwingendorf, Draga Vidakovic, Ed Dubinsky, Mazen Shahin, G. Joseph Wimbish, Jr.*

38. Models That Work: Case Studies in Effective Undergraduate Mathematics Programs, *Alan C. Tucker*, Editor.
39. Calculus: The Dynamics of Change, *CUPM Subcommittee on Calculus Reform and the First Two Years, A. Wayne Roberts*, Editor.
40. Vita Mathematica: Historical Research and Integration with Teaching, *Ronald Calinger*, Editor.
41. Geometry Turned On: Dynamic Software in Learning, Teaching, and Research, *James R. King and Doris Schattschneider*, Editors.
42. Resources for Teaching Linear Algebra, *David Carlson, Charles R. Johnson, David C. Lay, A. Duane Porter, Ann E. Watkins, William Watkins*, Editors.
43. Student Assessment in Calculus: A Report of the NSF Working Group on Assessment in Calculus, *Alan Schoenfeld*, Editor.
44. Readings in Cooperative Learning for Undergraduate Mathematics, *Ed Dubinsky, David Mathews, and Barbara E. Reynolds*, Editors.
45. Confronting the Core Curriculum: Considering Change in the Undergraduate Mathematics Major, *John A. Dossey*, Editor.
46. Women in Mathematics: Scaling the Heights, *Deborah Nolan*, Editor.
47. Exemplary Programs in Introductory College Mathematics: Innovative Programs Using Technology, *Susan Lenker*, Editor.
48. Writing in the Teaching and Learning of Mathematics, *John Meier and Thomas Rishel*.
49. Assessment Practices in Undergraduate Mathematics, *Bonnie Gold*, Editor.
50. Revolutions in Differential Equations: Exploring ODEs with Modern Technology, *Michael J. Kallaher*, Editor.
51. Using History to Teach Mathematics: An International Perspective, *Victor J. Katz*, Editor.
52. Teaching Statistics: Resources for Undergraduate Instructors, *Thomas L. Moore*, Editor.
53. Geometry at Work: Papers in Applied Geometry, *Catherine A. Gorini*, Editor.
54. Teaching First: A Guide for New Mathematicians, *Thomas W. Rishel*.
55. Cooperative Learning in Undergraduate Mathematics: Issues That Matter and Strategies That Work, *Elizabeth C. Rogers, Barbara E. Reynolds, Neil A. Davidson, and Anthony D. Thomas*, Editors.
56. Changing Calculus: A Report on Evaluation Efforts and National Impact from 1988 to 1998, *Susan L. Ganter*.
57. Learning to Teach and Teaching to Learn Mathematics: Resources for Professional Development, *Matthew Delong and Dale Winter*.
58. Fractals, Graphics, and Mathematics Education, *Benoit Mandelbrot and Michael Frame*, Editors.
59. Linear Algebra Gems: Assets for Undergraduate Mathematics, *David Carlson, Charles R. Johnson, David C. Lay, and A. Duane Porter*, Editors.
60. Innovations in Teaching Abstract Algebra, *Allen C. Hibbard and Ellen J. Maycock*, Editors.
61. Changing Core Mathematics, *Chris Arney and Donald Small*, Editors.
62. Achieving Quantitative Literacy: An Urgent Challenge for Higher Education, *Lynn Arthur Steen*.
63. Women Who Love Mathematics: A Sourcebook of Significant Writings, *Miriam Cooney*, Editor.
64. Leading the Mathematical Sciences Department: A Resource for Chairs, *Tina H. Straley, Marcia P. Sward, and Jon W. Scott*, Editors.

Preface

This volume originated with materials created for Leading the Academic Department: A Workshop for Chairs of Departments of Mathematical Sciences, held in July 2002 in Towson, Maryland, and in June 2003 in Reston, Virginia. These workshops were part of the Mathematical Association of America's Professional Enhancement Program (PREP), funded in part by a grant from the National Science Foundation, Division of Undergraduate Education. The current volume contains papers over a wide variety of topics that will be of interest to chairs struggling with the many demands on the chair (or head) and the various roles that person must play.

The theme of the workshops and of this volume is leadership rather than management; for it is the quality of leadership that will make the difference between a successful chair and one who is mediocre. Many people take the job of chair thinking of it as their turn to just somehow steer the department through the next few years, but most soon realize that is not enough and that there are opportunities to make a real contribution to the institution, students, and department faculty. This book is intended as a resource for individual chairs and as a vehicle for discussion among chairs within an institution, at a professional gathering, for future PREP workshops for chairs, or for similar workshops at MAA Sectional Meetings. While the perspective is clearly that of the mathematics department, most of what is presented here will be of interest to chairs of other departments as well.

Each workshop featured a few plenary addresses and panels led by experienced administrators. However, the primary format of the Chairs' Workshops was discussion sessions. The discussions were based upon case studies and issues papers, written by the discussion leaders and based upon real experiences. The case study approach proved to be an effective way for the department chairs to consider leadership from many angles and in different situations. Through discussion, the group shared not only possible solutions for particular scenarios, but also their decision-making processes and the many issues that must be considered. Successful strategies for being a leader of a mathematical sciences department emerged.

In the second Leading the Academic Department workshop, discussion sessions of specific issues were added to the sessions devoted to case studies. The discussion leaders of the second year workshop wrote the papers pertaining to these issues in order to set the context and focus of the workshop discussions.

Organization of the book: The book has four parts: a discussion of leadership or the wisdom of practice; case studies; issues papers; and the appendices containing other resources.

The wisdom of practice

The first paper of this part is a synopsis of the discussion of leadership led by the workshop leaders, all current or former department chairs, in July 2002. The session came at the end of the workshop, but here the discussion is presented first so as to set the stage. The following subsections of the wisdom of practice are devoted to the presentations of college and university administrators, two former university presidents, provosts, vice provosts, deans, and a university legal counsel. They all discussed the relationship between the person in their position and the department chair and offered advice on how to be a successful department leader. The two university presidents, Brit Kirwan (now a university system chancellor) and Joan Leitzel, are well-known mathematicians who have led major universities. As mathematicians and former university presidents, they bring unique insights to the position of the chair as leader of the Department of Mathematical Sciences. The view of provosts, vice provosts, and deans were presented in panels in each of the two workshops. Each gave advice hewn from personal experience in working closely with chairs of departments. The paper in the chapter written by a university lawyer follows the case studies format.

About the case studies

The case studies presented here are based upon real situations that occurred on campuses in the United States. They were written by the discussion leaders of the first workshop, although they are not all cases personally experienced by this group. Names, places, and other details have been altered to assure confidentiality. You may mistakenly recognize a situation and think you can identify it with a particular workshop leader. One participant told me that he recognized a case as having emanated from his institution; however, I had written that case and it came from my former institution, not his. Thus, there is a great deal of universality to these issues. Similar situations can arise in many departments. I encourage you to use and learn from these vignettes, but not to try to play detective. Furthermore, the cases are not meant to be publicized as proof that no matter how bad things are in your department, things could be worse.

About the issues papers

The issues selected for these papers are those most often mentioned by chairs as the important issues of the day. The papers were primarily written by the discussion leaders of the 2003 workshop as conversations they might have with colleagues who are mathematics department chairs. They are not meant to be the definitive statements on the issues. They are intended to stimulate thought and discussion. While many issues are raised in these papers, not all are resolved. Some of the papers present personal experiences, both good and bad. All of the papers will give you things to think about and will help you work your way through these issues.

How to use the case studies and issues papers

There are no resolutions presented to the case studies. The purpose is not to study what someone else did. That person's approach may not be the best for you and your department. These case studies are intended to help you to think through how you would handle the same or a similar situation. Perhaps you already have experienced a similar problem. This is the time to reflect upon your solution and what else you might have done. Similarly, the issues papers are not intended to be the definitive analyses of the topics presented. They were intended by the writers to be conversations with their colleagues who are also department chairs.

Although you may read this book on your own and use it for self-reflection, you will get added benefit out of it by discussing cases and issues with other department chairs. You may discuss the case studies that you find most appropriate with a department chair either in your own or another institution. You may organize informal meetings of department chairs at your institution for discussion sessions of the cases or issues. Or, you may organize discussion sessions for chairs at your MAA Sectional Meeting. Lastly, you may wish to attend the next MAA PREP workshop for department chairs. Check MAA Online at www.maa.org, and then choose Professional Development. The PREP workshops will be listed with all the information you will need to apply.

Preface

If you choose to organize informal discussion sessions or hold a workshop, you will need to fit your program into whatever constraints you have. At the MAA Department Chairs Workshops, discussion sessions last one and one-half hours each, and there are approximately ten participants in each session. The number of cases or issues discussed in one session is between three and five. In the first session, most groups spent most of their time on the first case they chose. However, in later sessions the groups were more wary of getting bogged down on one case and were able to cover several in one session. Many other scenarios can work. For example, if you have the opportunity for several short meetings, you might choose one issue or case study for a single discussion session.

The appendices

The appendices contain the entire MAA report, Guidelines for Departments, and the first part of the MAA CUPM Curriculum Guide 2004. The full texts of these reports and the companion to the CUPM volume, the Curriculum Foundations Project, have been sent to all mathematical sciences departments in the U.S. The MAA has for many years produced both Department Guidelines and CUPM Recommendations. These documents are very useful for conducting a department review that is either a formal process required by your institution or an informal process initiated within the department. These reports can be used as standards for setting department goals and as supporting materials for appeals for resources and other forms of support from higher administration. Only the six major recommendations of the CUPM Guide are included here. These six recommendations set the stage for the entire report and are intended to give you a broad insight to what the entire Guide addresses. Unlike previous CUPM recommendations, the 2004 publication addresses all student populations of undergraduate mathematics courses. The Curriculum Foundations Project presents the voices within the mathematics communities and of our partner disciplines, relating their visions for the mathematics component of their programs, especially in the first two years. The bibliography contains information about other resources that you may find useful.

Academic administration is a profession in its own right. Often there is no advance preparation and no opportunity for development of the skills required, only on-the-job training. Furthermore, that training is self-taught. The PREP workshops, this volume, the resources listed in the bibliography, and workshops or discussion sessions you participate in, may help to fill that gap.

While not everyone will find administration to their liking, many do find it challenging but also rewarding and fun. Whether you are chair for a short time or this is the start of a new career path, I hope that these materials will help you over some of the stumbling blocks, give you comfort and support in finding that others have been down the same road, and make the whole experience a better one for both you and the department.

Tina Straley
Executive Director
Mathematical Association of America

Contents

Preface vii
Contributors xv
Introduction, *Tina H.Straley* xvii

Part 1: The Wisdom of Practice

Introduction 3

The Chairs' Perspective
Advice From Experienced Chairs, compiled by *Tina H Straley* 5

The Presidents' Perspective
A President Speaks to Mathematics Department Chairs, *Joan R. Leitzel* 11
Leading Mathematical Sciences Departments Toward Excellence, *William T. Kirwan* ... 17

The Perspective of Vice Presidents, Provosts and Deans
A View from the Vice President's Office, *Anita Solow* 23
A View from the Provost's Office, *Paul Gaston* 25
A View from the Dean's Office, *Cheryl Peters* 27
A View from the Dean's Office, *Jimmy Solomon* 29

The Perspective of Legal Counsel
Academic Administration and Liability, *Michael A. Anselmi* 33

Part 2: Case Studies

Introduction 41

Students
The Non-placing Placement Test 43
The Disruptive Student 44
A Graduate Student Paddling Against the Current 45
TA Trouble 45

Faculty: Dealing with Troublesome Individuals
Too Late to Change? 47
Socrates Too Far 48

Faculty: Part-Time and Temporary Faculty

Some are More Equal 51
Keeping Adjuncts On Board 52
Adjunct Faculty—Welcome to the U.S.A 53
Last Minute Adjunct Hire 54

Faculty: Personnel Action

The Contrary Mentors 55
At the Line of Battle 56
The Hazards of Multiple Reviews 56
A Hiring Paradox 57
No Place for Sara 58
I Have a Dream 59

Curriculum

Jump-Starting Curriculum Change 61
Revising the Major 62
The Departmental Divide 63

Leadership

Search Burnout 65
Coming in as Chair: Where are Your Loyalties? 66
Departmental Management 67
Growing Pains 68
Staffing Summer Courses: Life in the Crossroads 69

Campus Politics

Contract or Not to Contract 71
Dealing Across Departments 72
The Interdisciplinary Team 73
Dean vs. Chair 74
The Persistent Fund Raiser 75
Vision 21 76

Legal Concerns

A Troublesome Dismissal 79
Free Speech vs. Discriminatory Behavior 80

Part 3: Current Issues in Mathematical Sciences Departments

Introduction 83

Students

Undergraduate Student Recruitment, *Carl Cowen* 85
Undergraduate Student Recruitment, *Bruce Ebanks* 87

Graduate Student Enrollment, *Carl Cowen* 89

Curricula and Programs
Encouraging/Leading Curriculum Renewal, *Norma Agras* 91
Encouraging/Leading Curriculum Renewal, *Catherine Murphy* 93
Encouraging/Leading Curriculum Renewal, *Martha Siegel* 94
Undergraduate Major, *Carl Cowen* 96
Technology in the Classroom: Winning Faculty Buy-In, *Donna Beers* 98
Technology in the Classroom, *Michael Pearson* 99
Program Review, *Donna Beers* 101

Faculty Issues
Faculty Review, *Connie Campbell* 103
Merit Pay, *Bruce Ebanks* 105
Hiring and Firing Staff and Faculty, *Bruce Ebanks, Jimmy Solomon, and Tina Straley* 106
Mentoring Faculty, *John Conway* 110
Mentoring: The Chair's Role and Responsibility, *W. James Lewis* 112
Dependence On and Culturalization of Part-Time and Temporary Faculty, *Norma Agras* 114
Dependence On and Culturalization of Part-Time and Temporary Faculty, *Catherine Murphy* . . . 115
Dependence On and Culturalization of Part-Time and Temporary Faculty, *Donna Beers* 116
Dependence On and Culturalization of Part-Time and Temporary Faculty, *Connie Campbell* . . . 118
Resolving Conflict within the Department, *Norma Agras* 119
Resolving Conflict within the Department, *Connie Campbell* 120
Managing Conflict with the Dean, Provost and Other Departments, *Connie Campbell* 121
Managing Conflict with the Dean, Provost and Other Departments, *Jimmy Solomon* 122

Finances
External Funding, *Donna Beers* 123
External Funding, *Martha Siegel* 125
Budgets, *Catherine Murphy, Jimmy Solomon, and Tina Straley* 126

Bibliography 131

Appendices

Resources for Department Chairs 135
Introduction from *CUPM Curriculum Guide 2004* 139
Guidelines for Programs and Departments in Undergraduate Mathematical Sciences 159

Contributors

Perspectives Papers

Michael Anselmi, *University Counsel, Towson University*
Paul Gaston, *Kent State University*
William T. Kirwan, *Chancellor, University System of Maryland*
Joan R. Leitzel, *President Emerita, University of New Hampshire*
Cheryl Peters, *Houston Community College*
Jimmy L. Solomon, *Georgia Southern University*
Anita Solow, *Genworth Financial, formerly Randolph-Macon Woman's College*
Tina H. Straley, *Executive Director, Mathematical Association of America*

Case Studies

Donna Beers, *Simmons College*
Bonnie Gold, *Monmouth University*
Raymond Johnson, *University of Maryland, College Park*
W. James Lewis, *University of Nebraska*
Daniel Maki, *Indiana University*
Celestino Mendez, *Metropolitan State College of Denver*
Catherine Murphy, *Purdue University, Calumet*
Jon Scott, *Montgomery College*
Martha Siegel, *Towson University*
Jimmy Solomon, *Georgia Southern University*
Tina Straley, *Mathematical Association of America*

Issues Papers

Norma Agras, *Miami-Dade Community College, Wolfson Campus*
Donna Beers, *Simmons College*
Connie Campbell, *Millsaps College*
John Conway, *University of Tennessee*
Carl Cowen, *Indiana University-Purdue University Indianapolis*
Bruce Ebanks, *Mississippi State University*
W. James Lewis, *University of Nebraska*
Catherine Murphy, *Purdue University, Calumet*
Michael Pearson, *Mathematical Association of America*
Martha Siegel, *Towson University*
Jimmy Solomon, *Georgia Southern University*
Tina Straley, *Mathematical Association of America*

Introduction

Tina H. Straley

Executive Director
Mathematical Association of America

One morning I was on my way to work, just like every other morning, but this was not going to be like every other day. I knew that it was going to be different, but I did not know in what ways. This was my first day as department chair. I assumed the position as acting chair in the middle of the academic year. There was no fanfare of a new year's beginning; there were no start-of-the-year meetings with the department; there were no new directories that listed me as chair. The college had split our department into Mathematics and Computer Science. The former chair was now chair of the Computer Science Department. We still had one budget that we were to share for the rest of the year. We divided the secretarial staff between the two departments, but they shared the same common office. That office door still said "Mathematics and Computer Science," so no changes were apparent. I was in the same faculty office I had occupied before. There was no sign on the door that said that I was chair. I entered my office and wondered what, if anything, different would happen.

A colleague stopped by to see me that morning. He needed to know how to handle a situation with a student. What that situation was I no longer remember. But I remember my reaction. I asked him why he thought I would know any better than he would how to handle that situation; why did he think that I had better insight into deciding what to do than I did yesterday when we were in the exact same positions? His answer was, "Today you have the authority to make the decision." And so, I began to be chair.

Throughout the years I have spent in academic and organization administration, I have pondered the same questions about what it means to be a leader. One can learn to establish management procedures that are efficient and effective. While that is a necessary condition for leadership, it is not the essence of leadership. Figuring out that essence is not as easy as mastering the day-to-day operations. Leadership is an abstract concept. Like art, you know it when you see it. While not everyone can be an artist, everyone can develop successful leadership strategies. If you did not have an interest in leadership, you would not be in the position of department chair. Someone or some group chose you to be the department chair; so, you must have talent they recognized. One reason that many administrative posts in academe are given to mathematicians is that we have the advantage of thinking analytically and being careful problem solvers, and we can translate these abilities to other contexts. Thus, we mathematicians have something going for us from the start. As you use the resources in this book and referenced here, you should reflect on your own definition of department leadership and do so continually hereafter. Whatever leadership is, it is always evolving and you will inevitably find unfamiliar territory that will change whatever conceptions you have.

Often as department chair you have to be a leader among your colleagues and close friends. This requires wearing a different hat without jeopardizing the relationships you already have. Often department chair is a rotating position, and the person who is in that position will be returning to faculty status after a specified time. Thus, leadership as a department chair has special challenges. The most important position in academic administration is the department chair. The department chair is the pivotal position where needs of students, faculty, higher administration, and the public all come together. For those who wish to progress through an administrative career, it is the one step that should not be skipped. The chair must possess all the skills of a leader while deeply integrated into the heart of the academic world as a faculty member. It is in the chair that administration and the teaching and scholarly missions of the college or university best come together. The chair is at the front line for all constituencies: students, faculty, outside communities, other departments, and central administration—just to name a few.

The function of the chair is multi-faceted, but the two essential roles are to be the leader of the department within the department and to represent the department as its leader outside the department. If you do not assume both of these roles, who else will serve in these capacities for your department? Who else in the college/university is thinking about the future of that department, its faculty, and its students? Who is speaking for mathematics on your campus? Anyone who has led an academic department knows that it is not a position of power, but it is a position of influence. You can make a difference; you can have a tremendous impact. But you cannot do it on your own; you must lead your department on this journey. All good ideas must be owned by the department and carried out by the faculty. The ideal chair advances the department, not the person. An effective chair may not get the credit when things go well, but an ineffective chair will get the blame when things do not go well. You must be the visionary, manager, facilitator, mentor, problem solver, promoter, and advocate. You will probably not have the time to be the complete faculty member that you were. You now have a different role. You might miss teaching and research, but you will play an important role in making the teaching, scholarship, and service of your department better than it was before you took this position.

You are not alone in this undertaking. You have other department chairs in your institution and chairs of mathematics departments in other institutions as colleagues. Both can be wonderful resources, in different ways. What follows is wisdom and advice gathered from chairs of mathematics departments and administrators in all types of institutions across the country. You might take comfort in knowing that your situation is not unique. Materials and references in the appendices and bibliography can aid you in deciding your departments' needs and what it needs to do. These materials may assist you in articulating a vision, deciding priorities, and setting a path. Mostly what you will gain from this volume is the time spent in reflection about your role as chair and the kind of leader you want to be.

Part I

The Wisdom of Practice

Introduction

The wisdom of practice from the perspective of experienced department chairs is a compilation of the comments given by the nine department chairs who served as workshop leaders at the first MAA workshop for department chairs in 2002. The chairs responded to questions about their management style and the way they promoted a vision and gave leadership to their departments.

The wisdom of practice from the perspective of university presidents is taken from the keynote addresses of the two workshops. Each keynoter has been a University President and a mathematics department chair. William E. (Brit) Kirwan spoke at the 2002 workshop on the day he concluded his service as President of The Ohio State University to return to Maryland as Chancellor of the University of Maryland System. Brit spent much of his academic career in Maryland, including appointment as President of the University of Maryland. The keynote speaker for the 2003 workshop was Joan Leitzel, just retired as President of the University of New Hampshire. Joan spent most of her academic career at The Ohio State University before moving to the University of Nebraska at Lincoln as Provost and then to New Hampshire. Both Brit and Joan have been Leitzel Lecturers (named for Jim Leitzel, Joan's husband and partner) at MAA MathFests.

The wisdom of practice from the perspective of deans and provosts is taken from presentations at the June 2003 Department Chairs Workshop. The panel of deans and provosts addressed the issues every department chair should know. The panelists, like the participants, represented the diversity of US higher education. Jim Solomon, who recently retired as Dean of Science and Technology at Georgia Southern University, in Statesboro, GA, chaired the panel. Anita Solow, was Vice President of Academic Affairs and Dean of the College at Randolph-Macon Woman's College in Lynchburg, VA. Paul Gaston, whose academic area is English Literature, is Provost at Kent State University in Ohio. Cheryl Peters, also in English Literature, is Academic Dean at Houston Community College, a large, urban, two-year institution in Texas. The panelists addressed the very important and complex relationship between the department chair and the chair's supervisor.

The final paper in this section is by Michael Anselmi, legal counsel at Towson University in Towson, MD. Michael spoke at the June 2002 workshop and again at the Joint Mathematics Meetings in Baltimore in January 2003. His case study approach fits well with the format of the Chairs workshop and gives context to his advice.

What you will find in the following section is much agreement and a lot of good advice. As chair your relationship with the central administration is of vital importance, since that relationship will determine your success in securing resources, support, and acceptance for the needs and goals within the department. You must be an accomplished advocate outside the department in order to gain respect and confidence within the department.

Advice from Experienced Chairs

compiled by

Tina H. Straley[1]
*The Mathematical
Association of America*

This chapter is a synopsis of a discussion at the 2002 MAA Chairs' Workshop on the role of the department chair as the department leader. The names of panelists for this session are listed at the end of this section. The panelists were asked two questions: What is your management style? How do you formulate and implement a vision for the department? The panelists spoke to these questions, but also took the dialogue far beyond them. What emerged was the collective wisdom of the group on how to successfully chair a mathematical sciences department.

While there was not always complete agreement among the panelists and participants, there was consensus on most issues. Everyone agreed on the benefits of a department having an effective leader, and all agreed that leadership transcends efficient management.

We hope that this synopsis will provide new, as well as experienced, department chairs with guidance as you strive to lead your departments to higher levels of achievement in teaching, research, and service.

The Job Description

The job description of department chair or head varies among different types of institutions and among institutions within the same type. Some department chairs are in 100% administrative positions while others continue to meet their faculty expectations in teaching, research, and service. Some department chairs have no more than a one course teaching load per term, while others may have only a slightly reduced course load. Some department chairs are appointed for a term of indefinite length and are not planning to resume faculty duties in the near future; others have a three- to six-year term and must think about what they will be doing after being chair. Some department chairs at two-year colleges or small four-year colleges do not have budget and personnel responsibilities and function more as area coordinators than as administrators. Management may be concentrated in one person, the chair, or may be distributed among assistant positions or distributed through the department via a committee structure. The formality of the structure is usually a factor of the size of the department.

No matter which model is in place at your institution, it is critical that you serve as the department's representative to the administration and to other departments. If you do not speak for mathematics on campus, who will?

---

[1] The author wishes to thank Ann Trenk for her assistance in preparing this synopsis.

You must facilitate the creation and implementation of a vision for the department, or your department will stagnate and eventually become out of sync with the profession and with the university or college.

Handling the Daily "Hum"

One panelist described his management style as handling the daily "hum" of activity in the department. As chair, you must handle that hum efficiently and effectively. If work gets backed up, if faculty and students do not get answers, if phone calls are not returned, if reports are not submitted on time, if staff is not properly supervised, then the work of the department will bog down and no one will be satisfied. The opportunity for you to lead, and for the department to grow and prosper, will be seriously compromised.

Here are some words of wisdom for new and not-so-new chairs:

- *Be open.* Information is power, but it is more powerful when it is shared. Decisions based upon information that only you have will be suspect. It is better to have people disagree with your decisions knowing on what basis you made them than to disagree and suspect your motives.

- *View yourself as serving the faculty, not the other way around.* The faculty want you to succeed. You can go to them for support. The best scenario is a reciprocal relationship. If you view helping the faculty do their jobs as a priority of your job and give them whatever support you can, they will reciprocate and support you in your job.

- *Don't try to do everything yourself.* The hardest part of the job is finding time for the big ideas because there are so many details to which you must attend. You must trust others. If you give someone a job to do, let that person do it, even though it might not be done your way. If you find that you cannot work with another person's style, don't assign that type of task to that person again. But when you find the right person for the job, help, but don't be overly directive. If you have the luxury of having administrative assistants, either faculty or staff, use them effectively. Delegating routine tasks will give you the time to devote to leadership and to your own personal duties outside of chairing. An assistant works best if he or she has responsibility over a particular area of department management and has authority to match that responsibility.

- *Meet deadlines.* A common, yet highly undesirable, operational model is *planned crisis management*. That is, things are not worked on until they are due. Thus, you find yourself constantly working in a crisis mode trying to make deadlines. You are forced to put off those things that are not yet due, and get around to them only when they become crises. This style of management is a source of much of the stress of the job. Breaking out of this pattern takes time and commitment to work on things when they are *announced* rather than waiting until they are *due*.

- *Lead, but don't try to "manage," your faculty.* College professors are highly creative and independent thinkers. It's the nature of the profession. They can't be managed, only led. It is vital that you know what the faculty members are thinking and what their needs are. A good way to find out is to walk around the department and talk to people. Go to the offices of the faculty members, talk to them there, and find out what they care about. You will be seen as more collegial and open to their opinions. It's much easier to lead people where they want to go.

- *Be available.* Know what is going on by being available where faculty can easily find you rather than require faculty to seek you out in your office, which is more formal. One chair has lunch in the department lounge every day. People know they can find him there. This environment invites casual conversation; it is even more relaxed than in the faculty member's office. Faculty members have an opportunity to broach subjects and ideas that are not necessarily fully formed, and you get the chance to work with them on problem solving at a very early stage.

- *Find a way to say "yes."* Your first reaction may be to say "no" because there is not enough money or resources or time for something the faculty member wants to do. Your job is to find the money, resources, and time. It may take creativity, it may take reorganization, it may require that you pull in favors. But, usually either you have more money than you think or you can get additional money from somewhere else on campus (not just from the dean's budget). If you can tie a project or activity to a campus priority or to a project or program funded elsewhere on campus, you might find the new resources you need. Most departments have a large enough budget for the chair to have some discretionary money. A few victories will go a long way for you to make points with the faculty, make their lives better, and raise morale. If you are moving money around in your budget to meet a faculty member's need, you are affirming the work of that faculty member. Faculty particularly like to see campus resources directed to their department. It is affirmation that the work of the mathematics department is valued across campus. In either case, you are letting department members know that you are willing and able to go to bat for them and their ideas.

- *Be at peace with indirect success.* While you are chair, others will write the papers, run the undergraduate research seminar, advise the math club, get the teaching awards, receive grants, lead the faculty senate, and be department stars. Your success is in making good things happen in your department, for your department's students, for your faculty, and for the department as a whole. A successful department means a successful chair.

- *Maintain good lines of communication with other departments and with the administrative offices around campus.* You are the public relations officer for your department. Know the registrar, the director of admissions, the director of finance, and so on. Also know some of the people who work in these offices. Don't just communicate when you have a problem. Find time and reasons to compliment them for jobs well done and talk to them about positive ideas for improving operations, especially if there is something that you can do for them. This ensures better cooperation when you must bring them a complaint.

- *Pay attention to details.* Get it right the first time and you won't have to redo it. Sometimes the way things must be done doesn't make sense. You may have to do it anyway, but then work with people on campus to develop better procedures for the future.

- *Success is permanently plugging the holes.* Every department chair has made a list of things to do that day only to find none crossed off at the end of the day (or even the end of that week.) The day is often spent on totally unexpected demands. Often people will say that administration is mostly putting out fires. Another analogy is plugging holes in a dam. A hole opens and water pours through. You put one finger in that hole to keep the water back, at least for awhile. Then another hole allows water to burst through and you use a second finger, and then a third, and so on. Once you have used all your fingers, how do you plug the next hole to open? You have to let one hole go, hoping it has filled in sufficiently with silt and will not cause trouble, so you can plug the latest hole. Often you may hope a hole will stay plugged at least as long as you are in the job and it will be someone else's worry after that. Common examples are complaints about a particular faculty member's teaching or lack of research, salaries for your department being below the national norm, threats from other departments to teach their own mathematics, lack of funding for research or special projects, and shortage of faculty for the teaching load of the department. You may find temporary solutions for such problems, but they keep coming back. But you must remain vigilant. Don't ignore a problem because the immediate crisis has passed. Keep working towards a permanent solution. Success is permanently plugging the holes.

Creating a Vision for Your Department

All chairs and participants agreed that the vision must come from the department, not from the chair. You can make suggestions and lead the department to concentrate on important issues. But you cannot define the problem and present the solution and then expect the faculty to march along together implementing your plan. You must talk to the faculty and find out what they think is important. Department meetings are too short to work out vision and strategic directions for the department. Ideas need to be formed ahead of time in the hallways, faculty offices, and in committee meetings. You need time for ideas to filter through the department.

If there is an issue that is not on the department's radar screen, start with the issue and present facts about that issue, not just anecdotes. Present the issue informally at first. If there are department members who are interested in working on it, encourage them. Bring the issue to the whole department at a meeting after there has been informal discussion. Ask for volunteers. Then meet again with the smaller group where you can suggest solutions for them to consider among their own ideas. The committee or task force should take their suggestions back to a department meeting. You should expect this process to take time to reach consensus or at least majority approval. Once the department has come up with a course of action, you can advance their vision to the dean, the higher administration, funding sources, and other constituencies (other departments or community groups, for example). The approach suggested here allows you to have important input, but the solution belongs to the faculty, not you alone. You don't want to be a leader that is so far ahead of the pack that when you turn around to rally the troops, there is no one behind you.

Not only do you need to canvas your department members to find out what is important, but also you should be canvassing others in the administration and other departments. Be aware of your institution's strategic plan and how that is to be realized in your college or school. If your department's vision fits the institution's plan, you will find it much easier to get support and resources. If your department does not see how their vision fits in with the institutional vision, tweak the institution's plan and tweak the department's goals until the two fit, even if loosely. The department ultimately must decide on what course to take, but it is your job to guide them to a course of action that is consistent with the direction the college is going.

Keeping current through professional meetings is an excellent way to sense the pulse of the profession. That message is often a powerful tool for the department in dealing with the administration.

Making Good Decisions

If you have to make a decision privately, and you often do, make sure that you are prepared to defend it publicly, even if you never have to. This way you are giving yourself a litmus test for fairness. Decisions that you make in private include such things as recommendations for personnel action, personnel reviews, and salary decisions. Sometimes deciding who is chosen to teach a prime course, or at the prime time, or who gets the best office, can be among the most sensitive decisions you have to make. And don't forget parking. It is one of the hottest issues on campus! No decision can be taken lightly. Make your decisions with integrity and courage. Not all decisions can be based upon popularity of the decision. Be fair, be firm, and be decisive when you have to be. Problems don't go away. If unattended, they fester.

Don't judge yourself on every decision or even on every day. Think about what you have accomplished in the long term, in a year, over a three-year appointment, over a decade. This way you can concentrate on the big issues that you have dealt with and have confidence that you will deal with the ones to come. Focus on quality; don't do things just to do them; wait for the good ideas. Avoid dissipating your energy; focus on your strengths and your department's strengths. Minimize weaknesses. Choose your battles wisely and carefully. You cannot win them all. You will be more successful if you don't expend your time and efforts on an unwinnable position. If a contentious issue is important to the department, rethink it and reformulate your approach. If you are not

going to support a department idea, you must answer to the department and explain why not, or why not now, or why not this way. There will be many times when the faculty in your department are not in agreement. This is not the time to stand back hoping things will work out. Nor is it a time to take sides. It is your job to pull the differing points of view together. Use real arguments if one approach or both are not feasible or will not accomplish the desired goals. If it is not clear which approach will work, further investigation must be done. Perhaps each approach should be tried. Diplomacy is a large part of leadership.

While the decisions of a department chair should not be made on the basis of a popularity contest, neither should they be made in isolation from the department and the administration. Being the department chair may be a balancing act between different constituencies and between different faculty members, sometimes between the students and the faculty. These are situations that will try your tactfulness, fairness, and decisiveness. There is no magic formula. Each situation is different. Everyone wants to be heard and wants her/his opinions to be taken seriously and respected. Once you have listened and considered all positions and have been fair and equitable in your decision, be prepared to explain it. People are not as offended about being disagreed with as they are about not being listened to. An explanation will show that you did seriously consider all points of view. Sometimes just listening is the most effective course of action.

Doing Research While Serving as Chair

Being department chair is more demanding than most expect. Two panelists reported that they are in their offices seven days a week. One said that the university expects chairs to work sixty-five hours a week; other chairs say that is not far out of line. However, most chairs do find time to work on research and teaching and to maintain professional activities apart from being chair. Some institutions openly expect the chair to maintain a research agenda or to continue to teach a full load.

The panel strongly recommended staying active professionally, whether it is expected or not. Often chairs are active in department projects that result in educational grants. Chairs at research departments often maintain their research funding. How to do this varies among institutions and departments. Most panelists do their own work after hours, when the traffic ebbs and they can close their doors, or on weekends, which is why they are in the office seven days a week. One panelist had administrative office hours and closed her office door at other times. This chair had little release time and was expected to fulfill all the demands on faculty. Most chairs stay active in their professional associations and attend meetings, seminars, colloquia, and conferences. Most chairs have to put some of their professional activities on ice, but they stay involved enough to be able to return to these activities after being chair.

It's a Great Job!

Being a department chair is a wonderful opportunity. The decisions you make and the practices you put in place can have a much greater effect on the lives of faculty and students in your department than you could possibly have as a faculty member. Not all faculty like the practice of administration and some should not be chairs. But for those who do like it, this is a great job!

Panelists: Donna Beers, Simmons College; Bonnie Gold, Monmouth University; Raymond Johnson, University of Maryland, College Park; Jim Lewis, University of Nebraska; Daniel Maki, Indiana University; Celestino Mendez, Metropolitan State College of Denver; Catherine Murphy, Purdue University, Calumet; Arnold Ostebee, St. Olaf College; Jon Scott, Montgomery College; Martha Siegel, Towson University; Jimmy Solomon, Georgia Southern University; Tina Straley, Mathematical Association of America.

A President Speaks to Mathematics Department Chairs

Joan R. Leitzel
President Emerita
University of New Hampshire

It is a pleasure to talk with you about the role of chairpersons at universities and colleges. My own faculty experience was at public land grant institutions: Ohio State for a long time, and later at Nebraska and New Hampshire. I know you come from a great variety of institutions. Our variety contributes to the richness of higher education in this country; we have many ways to educate students. My hope is that you will be able to translate my experience and observations into something that is meaningful to you.

In my experience, mathematics department issues don't differ a great deal across types of institutions. We all work on similar issues because we are trying to accomplish many similar goals. One thing I know is that you are more important than you realize. Chairpersons in a college or university are among the most important individuals in the administrative chain, and they probably have the toughest jobs. Think about the decisions that are made at the departmental level: essentially all curricular decisions, degree requirements, most of the hiring decisions, and most of the promotion and tenure work. Indeed, most fundamental decisions that affect college and university quality are made at the departmental level. That is where the business of teaching and research is done; so strong leadership in chair positions is essential.

As the chair of a mathematics department, you lead a high profile department. That can be both a plus and a minus. On the positive side, you won't have any trouble getting people's attention. On the other hand, you have high visibility and are often a target for criticism. The spotlight comes with the job so you will want to work effectively with the dean, provost, and president. The more successful you can be in those relationships, the more successful your department will be.

As you know, when one changes offices, one's perspective changes, so I'm sure I think differently now than I did when I was the vice chair of the department at Ohio State. Then, if I recall, my perspective was that the University regarded Mathematics as a service department; that the administration used Mathematics as a cash cow; that the dean worried too much about complaints from students and parents; and that the department would never be all that it could be because of limited funding, heavy teaching loads, under-prepared students, conflicting goals among our own faculty, and weak leadership at critical points. I don't know if I ever said these things, but that was what I believed to be true.

Now that I am a little farther away, I see the situation of math departments somewhat differently, and this is what I would say to department chairs. Please give the president or provost or dean the ammunition they need to work with the public. These people interface with parents and stakeholders of all kinds, and they need to know what is going on in your department. They don't have time to dig out information, and often they don't even realize what they need to know. So, you must get information to them even when they don't ask for it. For example, if you are using teaching assistants, they need to know how the department supervises them. They need to know how long the placement test has been used and how reliable it is. They need to know what connections the department has outside the department or college, particularly if you have faculty working with faculty from other departments or with local businesses or with the schools. They need to know about student awards and accomplishments. They need to know the good things your department is doing. This information makes for good public relations and helps presidents and deans when they are interacting with alumni. You'll get points if you give them the kinds of information they can use in settings that are demanding for them.

Another thing I suggest, particularly at this time, is to find a way to work with K–12 education. K–12 needs to be a front burner issue for most colleges and universities. No Child Left Behind has sharpened the public's attention to these issues, so let your central administration know that the math department takes its role seriously. Maybe you have some bright high school student taking calculus on the campus, maybe you have a "math day" where you sponsor competitions. Whatever it is, be sure people know what you are doing.

Another piece of advice: don't fuss about the department's service role. It's just part of the landscape, part of being a mathematics department. In fact, there are some plusses if you do it right. One is that other departments care about their students' success in math courses, and they can be your allies. These are people who can support you if you need a couple more faculty positions (as long as the funds don't come directly out of their budgets). These are people who, in a general education review, are likely to press for a quantitative literacy requirement for all students. Remember, as burdensome as it may be to teach all the students, we do want every student's program to include good mathematics.

Here's another thought. It is very important for mathematics chairs to help shape institutional priorities and then find ways for the department to be indispensable in meeting them. If you don't know what the priorities are for your institution or if you sense they are just being handed down to you, something is wrong. You should be part of creating these priorities. I know you've had at least one conversation with your dean or provost before coming to this meeting. Hopefully this conversation gave you a clear sense of what the dean or provost believes the institutional priorities are. We can hope they are written down for the world to read, but if not, you need to tease them out. Then the assignment is to figure out how the department can function in support of the institutional priorities, because if you are not swimming with the institutional current, you're probably swimming against too strong a current. Since mathematics touches almost everything, you should be able to connect with some of the institutional priorities and still do the other things that are distinctive to mathematics. In the worse case, if you're convinced the institution either doesn't have clear priorities or has wrong priorities, you'll need to work with the dean and provost to undertake fairly serious strategic planning.

One further recommendation for math chairs—help shape the institutional budget model. If you don't know what the budget model is, then you don't know how revenue flows to the programs or how costs are distributed across programs, i.e., the algorithms for distribution of funds and costs. When I went to New Hampshire, the algorithm was last year, plus or minus a little, as if sometime in the past somebody got it right and each year all we need to do is make marginal adjustments. Building budget models is serious business for universities and colleges, and math faculty are better at it than most. I want a model with these kinds of characteristics: to be responsive to changes in departmental activity and institutional priorities, to encourage revenue growth and discourage costs, to provide for flexibility in the use of funds, and to support long range planning. My experience is that there's a great deal of benefit in understanding the institutional budget model and working to adjust it if it really is a problem for your department.

What's Ahead for Colleges And Universities

I'd like to look with you at some of what I see ahead for universities and colleges in our country. For most of our institutions—and you'll recognize this—there are increasingly loud calls for greater public accountability. On the one hand, most legislatures are reducing state support; at the same time the public is asking our institutions to do more and to demonstrate that funds are used effectively. Private institutions are also seeing lower revenue as their endowment pay-outs decrease. With the tuition increases that have followed, families, especially lower income families, are worried that they will no longer have access to higher education. Consequently, you can expect to be challenged on issues such as faculty workload, salary increases, and tenure. My sense is that mathematics departments need to develop ways to measure what they are achieving in order to say with clarity, "this is what we intend to achieve, this is how we will know we're successful, this is where we are today." We don't like such expectations very much at universities; traditionally we haven't worked this way. We believe we know quality when we see it. And yet, at this time, we need to convince others in the public arena that we know what we want to do, that we are having reasonable success doing it, and that we can demonstrate our success.

Federal funding is important to our institutions in both research and instruction. There is some understandable concern that current funds will begin to be directed too much toward national security issues, but NSF funding for both mathematics research and mathematics education have had significant increases this year. Although the jury may still be out on federal funding for research and instruction, it is clear that federal financial aid for students is not getting the increases we hoped for. Student support is likely to be an even bigger problem down the road. We can expect arguments during the reauthorization of the Higher Education Act in 2004 about the formula for campus-based aid and about the federal government's role in student financial aid.

Can we hope that state funding for higher education will be restored when the economy recovers? I'm assuming not, and I believe institutions should behave as if the reductions are permanent, as hard as that will be in many places. One consequence will be that most public institutions will begin to develop other revenue sources more vigorously.

Many institutions are anticipating some enrollment changes. In New England, 2007 is expected to be the year when the number of 18-year-olds starts to drop fairly substantially. You can find out what demographics suggest for your region. There may be population shifts, and there are also the issues of nontraditional students. Lifelong learning has become very real in the knowledge economy, and I expect those enrollments will likely continue to grow. In addition, everyone hopes to become more successful with students of color and increase minority enrollments. It may well be that your institution is anticipating a somewhat different mix of students over the next decade, and that will be a factor in mathematics department planning.

Competition for faculty is always an issue, of course, and in mathematics there is no reason to believe hiring will get much easier over the next few years. Making arrangements for international scholars and faculty from other countries could become even more difficult with the increased levels of security and the monitoring requirements now in place.

There are other priority issues for colleges and universities. Some of these may be front burner issues at your institution:

- Undergraduate education will have a stronger emphasis in the next decade, I believe. For many of your institutions, undergraduates may have always been the most important part of the enterprise, but at other places their importance diminished and will be restored.
- Educating all sectors of the population remains a critical issue nationally. How do we set expectations for educating all the citizens and providing education in which they are successful?
- We have talked about K–12 education. I expect many colleges and universities will give more attention especially to teacher education and to math and science education.
- Don't be surprised if your institution attempts to narrow its focus in light of reduced funding and move toward more selective excellence. Usually, math departments can welcome this, anticipating that their programs have a preferred status.
- More partnerships are likely in the future, between programs, between institutions, with businesses, with government agencies.
- Interdisciplinary studies continue to be important with some softening of the current disciplinary boundaries.
- Expect your college or university to work harder to make stakeholders happy, especially those that have some authority over the institution and those that have money.

Among the things I've mentioned, you may find several priorities of your own institution. The important thing is to be clear about what those priorities are at this time and to educate your faculty about them. Then they can find ways to be supportive of institutional priorities as they develop departmental programs.

What's Ahead for Mathematics Departments?

I'd like to move from our consideration of what's ahead for universities and colleges to speculate now about the next few years for mathematics departments. We've talked about the requirements to document student performance and faculty activity, and to demonstrate that the department is doing what it says it will do and doing that with a high level of success.

The pressure from business and industry for more graduates in mathematics intensive fields is likely to continue, even increase. Science and engineering jobs in this country are growing three times faster than jobs generally. There was a 159% increase in technical jobs from 1980 to 2000, with another 50% increase projected by the year 2020.

More technically trained graduates means more mathematics enrollments.

What about the numbers of majors? We've gone from 16,500 bachelor's degrees in mathematics in 1986 to fewer than 12,000 in 2000. That's more than a 25% drop and can be a problem for chairs who want to retain the size of their faculty. Our doctorates have come down a bit (1,249 in 1971 and 1106 in 2000) but more sobering than the count is that almost half of our doctorates—47% in 2000—are earned by international students, many of whom now return to their home countries. In addition, the international student situation could change over the next decade with the maturing of the EU. Reciprocity among the universities in EU countries means that German students can go to France or Hungarian students to Germany at costs substantially less than coming to the U.S. I expect to see a change in our enrollments from Europe, and hopefully an increase in numbers from Africa and Latin America.

Technology is an institutional issue, but one that affects math departments especially. Technology is changing not only how we teach, but who we teach and what we teach. We have amazing tools, and faculty are finding effective ways to use them in the learning of mathematics. Initially, many institutions bought hardware with one-time money and may not have budgeted adequately for staffing, maintenance, replacement, or even space. Twenty years ago some naïve folks thought technology would be a big budget savings for our programs! Not true, at least in the short run. It will be costly and must be budgeted, not patched in.

We can expect to see more mathematics requirements for undergraduates in several fields, for example, the life sciences. It used to be that a student in the life sciences would take minimum mathematics, but those requirements are developing to be not only more mathematics but somewhat different mathematics. At many schools quantitative literacy now has the status of composition, reflecting the recognition that every student must have certain mathematics tools and understanding. If you haven't had those conversations yet on your campus, you likely will in the near future. It will be important to address curricular changes by rethinking the entire curriculum and not just by adding on more courses.

We still have some math-traumatized students coming into our institutions. There are predictions that high stakes testing in K–12 could produce more. In any case, we don't know what effect the mandated testing of every child, every year in grades 3-8, will have on curricula or on student learning. Within a decade you could see changes in student preparation in mathematics, for better or for worse.

As a reaction to budget reductions, we are seeing increases in temporary and part time adjuncts in math departments. These instructors can teach skills courses, they can bring the outside world into the classroom. But they can't shape the curriculum, they can't advise students, and they can't carry long-term departmental responsibility. I regard adding part timers as a short-term response to problems that weren't fully anticipated. Even three-year lectureships can be destructive to people's careers if they are not well-structured. I hope chairs and deans will be able to find different kinds of responses to budget problems. Down the road, too many short-term appointments will come at a cost—possibly a higher cost than doing the hard planning now to find alternatives. deans, provosts, and presidents can be influenced by information about what is happening at other institutions. Your institution compares itself to some set of peer institutions. You will want to know what the mathematics departments are doing in those institutions so that when you sit down with your dean, you can tell him/her what is happening in comparable schools.

I'd like to return again to issues around funding. (If you're now thinking that presidents spend almost all of their time focused on funding, you're approximately correct.) There are several sources of institutional funding, some of which work particularly well for Mathematics. You wish you could keep the tuition from the enrollments in math courses. That's a budget model issue; as I mentioned before, you'll want to pay attention to the model. For public institutions there is state subsidy and often, also, state grants. Private institutions that don't get the subsidy may, however, be able to compete for certain state grants. You should become well acquainted with programs in the federal agencies. Private foundations are somewhat more difficult to navigate, but probably someone at your place knows how. Right now industry is having trouble keeping their own organizations afloat, but in better times they can be good partners. Also, don't hesitate to partner with math departments in other institutions. If they have something to offer and you have a complement, together you may be able to shape a program or a proposal. And keep in mind gifts from individual donors; occasionally there is one who especially likes mathematics and recognizes its importance.

Summary

Finally, let me summarize some of the roles I hope you'll play as department chairs.

- It's your job to clarify the department's vision and to separate the long-term goals from the short-term goals. Sometimes a goal is long-term because it will take a long time to get there, but other times a goal is set in the future because it just can't be done in the current circumstance. There are barriers that can block a particular goal for a time but will disappear eventually, so be realistic in choosing the timing.
- Understand the priorities of your institution and develop a departmental plan that is compatible with them.
- Use the institutional budget model and budget processes to support departmental priorities.
- Hire well and nurture the junior faculty; they are the future.
- Give special attention to the undergraduates; while we can expect all faculty to do this, the chair needs to be especially aware of the undergraduates.
- Position the department for accountability by setting concrete measures of success.
- Market your department internally and externally.
- Recognize both departmental and individual successes and celebrate them.

Being a mathematics department chair is not for sissies, but it can be rewarding. You will have many opportunities to better your institution, your department, and most of all, your students. Good luck to you all.

Leading Mathematical Sciences Departments Toward Excellence

Excerpts from an address by

William T. Kirwan
*Chancellor,
University of Maryland System
Former President,
University of Maryland, College Park
and The Ohio State University*

Before I begin, I want to offer a disclaimer. As much as I still love mathematics and the company of mathematicians, I have long been removed from the world you live and work in on a daily basis. I did serve as a teaching assistant this past term in a self-paced, computer-based calculus course, and it was a wonderful experience. Indeed, the 50 minutes I spent in the classroom was often the highlight of my day. But that experience notwithstanding, I fear that any substantive observations I might make are things those of you on the front line have already thought about…and either dismissed or implemented.

Let me start by congratulating the organizers of this very impressive MAA workshop and the Task Force that prepared the book *Toward Excellence: Leading a Doctoral Mathematics Department in the 21st Century* [13] I am deeply impressed by the book's content and the directions it sets. I believe there is much wisdom in it for chairs of other disciplines as well, and I hope it will be widely circulated throughout academia, not just in mathematics departments.

If I have a quibble with the book, it's with its opening premise, which suggests that the sole objective is to fill the coffers of mathematics departments. Now obviously that's an important pragmatic objective for any department chair worth her or his salt. I do wish, however, that the book had at least posited the notion that its recommendations were important because of their intrinsic merit…and that if we do these things and do them well, there is reason to believe that our budgets will grow. This quibble aside, I think the advice and the examples of best practices found in the book are squarely on the mark.

Toward Excellence is a revolutionary document. I was pleased—surprised, but pleased—by the book's sense of urgency, one might even say passion, about the need for much greater attention to undergraduates and the undergraduate curriculum. The plan of action it offers leads to a radically different department than the kind in which I was educated and which I inhabited during my active days as a mathematician. When I began my career in the mid-1960's, some colleagues bragged that they prepared for undergraduate classes while walking from their offices to the lecture halls. If the truth be known, on occasion I was probably among those braggarts.

What has caused this shift? Why in only a few short decades have we moved from a culture where undergraduate education was almost an afterthought to one which issues a clarion call for a new departmental paradigm, one that places undergraduate education at the very center of departmental concerns? And are the causes temporal, fads that will burn brightly for a few years and then fizzle, heralding a return to the "good old days?"

My unqualified response to both questions is an emphatic No! This change reflects fundamental societal changes, driven in part by economic forces that have radically changed our nation's socio-economic ethos. As creations of society, our universities—and higher education generally—cannot help but be influenced by the forces at work in the larger societal context within which we operate. As always, these forces will change over time, but the change will be evolutionary and on the time scale of decades, not years.

Those few of you, relatively speaking, who grew up professionally in the Sixties, as I did, will recall a fervency, if not outright paranoia, about our competitive position vis a vis the Soviet Union. For those of you too young to recall the event, it is difficult to convey the impact that Sputnik had on the American psyche. With the launch of this Soviet space vehicle, our nation became convinced that we had lost our scientific dominance. This was unacceptable to the American people, and enormous sums were invested to reestablish our scientific preeminence.

Maybe some of you were beneficiaries, as I was, of the National Defense Education Act, which gave thousands of Americans the opportunity to get a Ph.D. fully funded by the federal government. The sine qua non of national policy leaders, indeed the entire body politic back then, was to build the research capabilities of our major universities and national laboratories. Fortunately, the nation either had or borrowed the money necessary to press forward with this goal.

All this led to a university culture where:

- Research was the central professional endeavor and focus of academic life;
- The quality of professional activity was determined almost exclusively by external formal peer review;
- Knowledge could be pursued for its own sake, usually amply supported by the federal government;
- Research was discipline-based, with greater and greater emphasis on increased subspecialities;
- Faculty loyalty was directed away from the university and toward disciplines and subspecialties.

To be honest, my generation grew up in an era and with a set of priorities unlike anything higher education had known in its previous history and is unlikely ever to see again.

But in addition, other *social* dynamics were at work. The Sixties was an era of large government programs aimed at building Lyndon Johnson's Great Society. It was the era when, for good reason, affirmative action reached full flower and wide acceptance. It was an era of government supported and government sponsored entitlements, large, well-funded social programs and rapidly rising government spending. This was the larger socio-economic dynamic that guided the development of our universities through the Sixties and Seventies and well into the Eighties.

I make no claim to being a social historian, but I think the forces I just described, losing energy through the Eighties, met their demise in the nation's economic downturn at the beginning of this decade. Ironically, the final blow was a development with a "Sputnik-like" effect. For while our economy was in near free fall, the Japanese economy was booming—a replay of our scientific competition with the Russians. Experts began writing that Western economic dominance, and most especially U.S. economic dominance, had come to an end. A new economic order had replaced it, they said, led by Japan and emerging nations on the Pacific Rim.

Some of you may have read *The Reckoning*, David Halberstam's chronicle of the rise of the Japanese auto industry and the concurrent decline of Detroit. And you may recall a PBS series in the early Nineties in which leading economists described the new economic phenomenon. To add insult to injury, the Japanese started buying up some of our most treasured symbols of power—major office complexes in New York like the AT&T headquarters and Rockefeller Center, and perhaps the cruelest blow of all, the Pebble Beach golf course. Talk about a challenge to our manhood!

The reaction was no less dramatic than with Sputnik, only this time the response was lead by the private sector, not the government. Focus, accountability, relevance and the bottom line became the new national mantra. Companies like IBM, which had long heralded their corporate version of academic tenure, began massive layoffs. But fortunately, America retained the technological prowess and the entrepreneurial culture to recapture the dominant position in the world economy. We soon regained our leadership position, in the process pioneering an information age economy that is leading us into a new century.

One corollary of our economic success is the new socio-economic culture that has infused the private sector and government at all levels. It's important that universities understand this new culture, for it is unlikely that any university can succeed unless it takes these new socio-economic dynamics into account.

I've taken the time to offer this little retrospective in order to emphasize two points. First, I think the larger social context within which we operate is fundamentally different than the one most of us grew up in...and that this context is not likely to change any time soon. Second, many faculty members do not seem to grasp this reality. In fact, I'm constantly amazed at how often faculty think that the changes universities are considering, and in some cases implementing, represent a sinister plot hatched by a devious coven of unenlightened presidents and provosts.

"If you would just tell the legislature how important what I'm doing is, they would get off our backs and give us more money," is a refrain I hear often.

Now the only thing more certain than change is resistance to change. Bob Chase, president of the National Education Association, described a bumper sticker given to him by a student. "Change is good," it read. "You go first."

Now I *don't* mean to imply that we are merely pawns of forces we cannot influence. Clearly, we must actively and forcefully explain the importance of our fundamental mission to policy leaders and the general public. We must openly champion the value of a liberal as well as a professional education and the dangers inherent in under-funding the arts and humanities. We must be strong advocates for the essential role of basic research. And along with enlightened trustees and other friends, we can and should significantly influence the public policy decisions that impact our universities.

But we must also understand—and work within—the present socio-political context of the larger society because it, too, will greatly influence our success.

What are some of the implications of this new order for higher education? The good news is that the body politic sees much of what we do as vital to our nation's continued well-being. Never in my lifetime has there been such a pervasive focus on the quality of education at all levels. In particular, the light has dawned on the private sector, which now recognizes that their success, unlike that in earlier economic booms, depends upon our producing an abundance of well-supported research efforts, at least when people see their connection to our economy.

The *bad* news? There is an aversion to any increase in taxes and to growth in government. This suggests that we can anticipate the following three trends in the coming years.

First, we will realize only modest increases in state funding, except in selected areas—primarily technology related—where there is a clear and obvious connection to economic growth. Targeted funding for programs that show promise of improving the quality of K–12 public education is likewise possible.

This means we will face continued financial austerity because, in the years ahead, it will be difficult to augment modest state increments with significantly increased student tuition. Many universities have already played the tuition card, and further significant increases conflict with society's expectations for access to higher education and workforce development.

A second phenomenon that will affect our lives is a continuing demand for accountability. Now, I am a strong proponent of accountability and am working hard to increase accountability at Ohio State. But any good concept can be carried too far, or be misused, especially when it comes encumbered with time-consuming processes that rely on simplistic notions of performance. But however burdensome this trend may become, the push for ever greater accountability is probably an inevitable consequence of tight state budgets.

With increasing competition for a limited pool of resources, every allocation will receive heightened scrutiny. Governors and legislators will feel forced to document the impact of state funds. The push for faculty workload measures, monitoring of retention and graduation rates, measurement of time-to-degree, reporting of course availability, and calls for measuring the "value added" of a college degree will continue. And there's no place to turn for relief from these demands because the general public thinks they are reasonable things to measure.

A third trend is an internal response to declining resources and increased demands for our services, namely the phenomenon of decentralized or responsibility-centered budgeting. While this phenomenon has existed within *private* universities for decades, if not since their inception, it is a much more recent phenomenon within *public* universities.

Simply put, it is a strategy to more tightly align a university's revenues with its expenses. Thus, research overhead and tuition generated by a college or department from its enrollments would be returned to the generating unit, minus some tax to pay for central services, such as the library, the registrar's office, etc. I started to say the president's office, but I'm not sure anyone would recognize that it provides a service.

Theoretically, such a system provides the incentive for departments to do things that are in demand and can generate new revenue. And while to a certain extent that does seem to occur, this system also has obvious downsides. For example, it creates a *dis*incentive for colleges to encourage their majors to take courses in other colleges, since presumably this represents a lost revenue opportunity. But whatever the merits, universities across the country, including my own, are implementing these decentralized budgeting systems.

Given these three trends—limited budgets with funds targeted for perceived societal needs, increased accountability for state funds and responsibility centered budgeting—let's consider how today's mathematics departments might fare.

Even I was surprised at how bad the data are: from 1991 to 1997, there was a significant decline in overall enrollment and a vanishing number of majors; a precipitous drop in graduate enrollment; and over sixty percent of course instruction was at the first-year calculus or lower level. Yet, there was essentially no decline in the number of tenured faculty, and there was an actual rise in the number of non-tenured teaching faculty. In the environment I've described, it's impossible to imagine that these trends can continue much longer. Either enrollments will go up or faculty positions will go down.

But the situation isn't *all* bad. Mathematics has a number of impressive assets. Along with English, it is still seen as one of the two core disciplines upon which an educational foundation is built. As a result, math is still a requirement at most colleges and universities. It always has had and undoubtedly always will have strong connections to other disciplines, especially in the physical sciences and technology and increasingly in the biological and social sciences. These connections occur, of course, at both the educational and research levels. And mathematics is singled out with special importance in the move to improve the quality of K–12 education.

In thinking about the issues you are facing, I tried to imagine myself as one of you, chair of a mathematics department. I asked myself, What would I do? How would I begin to address these problems?

Here's what I *think* I would do. First, I'd prepare for and then call a department-wide meeting, and invite the dean to give his or her perspectives on funding priorities. I'd share enrollment data and trends. In sum, I'd try to create a sense of urgency for change.

Hopefully, such a session would stimulate creation of a task force on the future of the department. I'd also make certain the dean and the provost were informed and supportive of this effort. I then would ask myself if I were the dean, provost or president, what would I like to see contained in the task force report?

First, I'd like to see a commitment to the centrality of undergraduate education in the department's mission. I'd like to see a call for a reshaping and restructuring of the curriculum with

- greater emphasis on active learning at all instructional levels,
- a call for departmental leadership in addressing the university's retention issues,
- a commitment to the development of joint majors and upper-level service courses in partnership with other departments, and,
- the creation of a departmental tract to prepare K–12 teachers.

Next, I'd want to see a commitment from the department to assume leadership in the creation of a statewide consortium on K–12 issues. This consortium would include representatives from other mathematics departments in the state, high school math coordinators and teachers. It would address mathematics expectations of the universities' matriculates, discuss curriculum issues and develop mechanisms to provide feedback to local schools on their students' preparation and performance.

I'd also like to see a plan from the department on its future faculty recruitment strategies. The plan would indicate how positions would be reallocated to hire people in interdisciplinary fields such as the computational sciences, neural networks, and string theory, as well as experts in the teaching and learning of mathematics.

I'd want the plan to address the graduate curriculum as well, how the department intends to change the curriculum to reflect the changing employment circumstances and the needs of the discipline. I'd want to see a plan that described a graduate program that provided non-academic career opportunities *and* prepared future faculty to be good stewards of an undergraduate program as well as able researchers.

Finally, I'd include a commitment to implement individualized faculty workloads. I think there is no area where universities' actions are more subject to valid criticism than the way we utilize our human resources. I'm certain the situation has improved somewhat from when I was a chair, but surely not enough. In my day, we had a uniform teaching load regardless of whether faculty were actually engaged in research. Indeed, we perpetuated a myth that *everyone* was involved in research. God help the person who tried to take on some special assignment involving undergraduates. Rumors would start to buzz in the hall. What's wrong with old Joe? His career must surely be on the skids. Such attitudes were never justified, but with the issues facing our departments and universities, they can no longer be tolerated.

We *do* need a new reward structure, one more in tune with current reality, and presidents, provosts and deans alike need to lead and ease the way for their creation. Mathematics departments are wonderfully suited to help support this effort. The demands on the department are perhaps more varied than any other department in the university and, given the data in the book, it's impossible not to conclude that personnel in most departments are under utilized. I'm confident that a department coming forth with a creative strategy in this area would gain enormous favor with the administration. Certainly, they would with me.

In advocating individualized work loads, I want to make one point very clear. At research universities, we have every reason to expect that candidates for tenure will demonstrate a research mastery of his or her field. In mathematics, this would almost certainly mean research published in major peer-reviewed journals. But, as Ernie Boyer pointed out in his brilliant essay, *Scholarship Reconsidered*, careers in academia are long, and few sustain an uninterrupted, 40-plus-year career of important research activity. As we all know, this is especially true in mathematics. Let's accept and take advantage of this fact. Let's make it not only possible but admirable for people to contribute to the department's many responsibilities. Let's make it possible for individuals to gain promotion to full professor if they excel in scholarship related to the learning of mathematics or in outreach to the K–12 sector.

Enough of my preaching on this topic. As you can sense, it is one on which I have strong views.

In any event, once a plan is developed, I'd invite the dean, provost and president in to hear a presentation on it and seek their support for its implementation. I can't speak for any other president, but I can tell you such a plan would get my attention and support. I'd want to use it as a model for other departments. And I would certainly want to insure that any department willing to commit to such a plan had the resources to do so.

In conclusion, one final word of perspective. When we talk about change of the magnitude under discussion at this conference, one can be left with the impression that everything being done today is bad. It is true that our rhetoric does tend to get away from us so let me state for the record that I don't feel that way. Many good and valuable things are under way. We don't need to change everything.

Certainly, chairs of departments at research universities have an important responsibility to sustain the high level of research that has produced remarkable advances in our discipline. The identification and nurture of bright young talent, the support of established, productive researchers is a vitally important responsibility of a department chair. Any implication that this is not the case is, in my view, terribly misguided and is most definitely *not* what I'm suggesting.

What is needed so desperately at this time is balance. We need balance in the department's mission, reflecting the multiple responsibilities we have. We need balance in the allocation of resources to meet these multiple demands. And most of all, we need balance in the allocation of people's time and responsibilities.

This is not an easy time to be a department chair, especially chair of a mathematics department. But, I'm confident that if you can help effect the kind of changes in your departments that are being called for, it will be one of the most rewarding experiences of your lives and among the greatest contributions of your careers.

A View from the Vice President's Office

Anita Solow
*Genworth Financial,
Formerly Vice President for Academic
Affairs and Dean of the College,
Randolph-Macon Woman's College,
Lynchburg, VA*

What I expect from department chairs

(other than "faster than a speeding bullet, leaps tall buildings in a single bound")

My first expectations of a department chair are honesty, mutual trust, and respect. Without these qualities, no one is going to get very far in administration. I have met department chairs who are not forthcoming with the full truth, and it is not very productive. I have met department chairs who are conspiracy theorists; they are so busy distrusting the administration that they are ineffective in dealing with that administration.

Next on my list is timeliness. Do not hide problems until they have grown so huge that you cannot hide them any longer. Giving the dean a "heads up" early on, before you need to take the problem to the dean for assistance, is a good strategy. Then there are no surprises when you do seek help, and you have already shown that you tried to deal with the situation on your own. On the other side of the coin, don't rush into the dean's office either. Before taking a complaint or other situation to the dean, do some research and try to resolve it on your own. Come to the dean with possible solutions, not just the problems. Have a back-up plan, not just one course of action. If your course of action is turned down or not feasible, you have another route to suggest.

The other aspect to timeliness is to complete reports as requested; these include annual reports, assessments, budgets, schedules, and so on.

Choose carefully what matters to take to the dean. If you are silent about all problems, then you are on your own when help was possible. If you complain all the time, your supervisors will stop listening and nothing will happen.

Be a solid advocate for the department. However, and this one is harder, have an institutional viewpoint at least some of the time. Sometimes you have to be a good citizen even though it is not in the best interest of the department. It does not hurt to gain a few brownie points with the administration and a few allies among the other department chairs. This duality of qualities makes you a good team player, makes you flexible, and lends credence to you when you must be insistent and will not bend.

As chair you must mentor your faculty, both regular faculty and adjuncts. As supervisor you are both advisor and judge. While you are helping new faculty adjust and senior faculty change and grow throughout their careers, you are at the same time assessing and reviewing them. You should pass on knowledge you have to the faculty member of both good and bad performance. Make sure you document problems as part of the review process so that the person gets continual feedback and has a chance to respond before it is too late. Do not write glowing reviews and then tell the dean about problems with or complaints about the person. Sometimes you have to take on the hard job of telling a person that there are problems; you cannot pass the buck to the dean to deal with the problems while you are the always the nice guy. Remember, you have to live with the faculty member; the dean doesn't. The dean cannot be the one to initiate the conversation about problems when the department chair has never broached them. That not only puts the dean in the role of heavy, it makes it look like it is the dean who has the problem, not the faculty member who is doing very nicely within the department. This leads to suspicions that the dean just doesn't like or understand mathematics (or whatever other discipline) faculty.

In all personnel matters (searches, promotions, merit) follow the law to the best of your ability and ask, if you are not sure, before you act. In fact, ask even if you are pretty sure.

I expect chairs to be knowledgeable about best practices and new ideas in their own disciplines. For example, in mathematics the chair should know the MAA Guidelines for Programs and Departments. The chair should be open to change. Being knowledgeable about trends and standards in their disciplines will help establish parameters for change in their departments.

The chair will have to make difficult decisions about staffing, curriculum, etc. Let me give you an example. Each department chair is asked to submit the course schedule for his/her department. Good chairs work with the members of the department to assure that certain courses are not scheduled at the same time (both within the department and with some courses outside of the department), and that there is a variety of time slots used. Poor chairs ask each member of the department what they wish to teach and when and then just put that down on the schedule form. I had a department chair (not mathematics) who scheduled all the course sections on Tuesday/Thursday and scheduled all the advanced courses at the same time. It is easier to just do what each faculty member wants but it is not in the best interest of the students or the institution.

Make the most of the position. At my institution, department chair is a three-year rotating position, renewable once. Many view it as a minimalist job—do the least required and maintain the status quo. But the most successful chairs are the ones who lead their departments, who ask questions, and who try to move their departments forward. Then one must leave gracefully. Make sure to pass on reports and information to the next chair so that each chair does not need to start at ground zero.

What department chairs should expect from the dean or provost

(other than having a hidden cache of money).

First and foremost, the chair should be able to go to the dean for support for the department. However, you will need solid evidence that the support is needed. You will often need external verification that requested resources are necessary.

Expect the dean to ask hard questions. This is necessary because the dean will have to present your requests at the next level and she will need the answers to those hard questions. You depend on the dean to know what answers she will need.

Expect decisiveness. It is important to get timely answers to your requests even if you do not like the answers. However, do not expect anyone to make a decision on incomplete information. You should not do this in the department; and the dean should not do it for you. Going forward before you are ready will only hold things up. The dean will likely only promise to investigate and get back to you and that may be with a request for more information. But again, it is important for the dean to respond in a timely manner.

The dean should be a resource for answers to legal problems. I unfortunately, confer with lawyers on a regular basis, but I do this to assure that we are acting within the legal bounds.

The dean should give you as little busy-work as possible. The dean should think through requests to the chairs and not fire off a memo for a report for every question that comes across the dean's desk.

Dealing with the mathematics department
(as a mathematician)

I find it harder to oversee the mathematics department than with other departments because I really do know the discipline and its trends. When I feel the mathematics department is not all it could be, it is doubly frustrating to me. With other departments I am not an expert nor do I pretend to be one. However, I have to remember that I am the dean and not the chair, and I cannot be more prescriptive with mathematics than I am with other disciplines. I tread a fine line between advice, direction, and being supportive of the department and its goals and desires.

A View from the Provost's Office

Paul Gaston
Provost
Kent State University, Kent, Ohio

It is indeed a pleasure to talk with mathematics chairs representing a broad diversity of experience, of institution, of longevity, and of geography. But it is also daunting, as a quick review of some of your introductions (and the concerns listed there) suggest. I will begin our conversation with four broad axioms—and will hope for your responses in return.

The first sounds simple, but represents a challenge: remain—or become—an advocate of mathematics for all students. That is, share with me the view that computational knowledge represents an essential component of a college education. That does not necessarily mean that all students must learn calculus—or even algebra. It does mean that all students, even those who have taken algebra or the Calculus, should become, in the word popularized by John Allen Paulos, "numerate," capable of understanding and applying mathematics in their disciplines, in their professions, and in the world. When Ivy League graduates are unable to describe in meaningful terms the order of magnitude separating a million from a billion, we face a challenge of numeracy.

The second also sounds simple, but it is also deceptive: commit to learning. Not to teaching, no matter how inspired, however well supported by technology. Not to courses, with their hour requirements, however well justified they may be. Not to degree programs, with their course requirements. While all of these are important, I speak instead of a commitment to students actually getting it, actually learning mathematics, by one means or another, whatever is required. Good teaching, solid courses, and coherent programs are all critical, but they must never be regarded as ends in themselves. They are means to that elusive end, the moment when the student stops scratching her head, smiles, and says, "I've got it."

My third suggestion is that mathematics department heads and chairs acknowledge that it takes a university to make a student competent in mathematics—and that they act on the expectation that their colleagues in other departments will do their part of the job. I am not necessarily speaking of "mathematics across the curriculum," although some institutions have found that approach a valuable one. I am speaking of the responsibility all faculty members share to face squarely the computational and statistical elements of their disciplines —from history to journalism to biology—and that they require of themselves the mathematical competence necessary to teach their students these elements, even if that means a refresher course or two for them with their colleagues in mathematics.

My fourth suggestion is that you attach a significant

value to the professional responsibility you exercise through your chairmanship or headship. It is traditional to express begrudging acceptance of administrative burdens, to complain that we would much prefer to be teaching and doing research, to explain that we are but serving until someone else might be victimized. We may enjoy that language from time to time, but we must take seriously the charge that we have been given. For the term of your leadership, you will exert to one extent or another a profound influence on the lives of your colleagues and the educational growth of hundreds, perhaps thousands, of students.

Of course, you would not be at this workshop if you were not serious about the obligations and opportunities that a chair enjoys. But your earnestness will take you only so far. When a student files a grade appeal, when a respected faculty member throws a tantrum in your office, when the dean weighs in too heavily on a tenure decision, or when the college curriculum committee vetoes your department's thoughtful reform proposal, you cannot rely only on your academic values, your native integrity, and your common sense. An awareness of technique, an acquaintance with best practices, and at least a rudimentary knowledge of the law of higher education may become critical. And you are here in part to weigh techniques, to learn best practices, and to gain some awareness of procedural expectations.

Fortunately, none of these suggestions is exactly revolutionary, yet in my experience they are sometimes honored more in the breach than in the observance. The symptoms of breach are familiar: a high-minded insistence on "standards" when students fail to gain a mathematics toe-hold; a shuffling of the curriculum or of class size or of technical aids when thoughtful attention to course objectives and pedagogy would accomplish more; or a rigid insistence on College Algebra as the narrow gate through which every aspiring baccalaureate must pass. None of these positions is without some merit, and an insistence on integrity in the curriculum cannot be all bad. But they can be symptomatic of an unwillingness to entertain the suggestions I have offered:

- a commitment to essential mathematical competence for all;
- a focus on student learning;
- a willingness to enlist colleagues from all disciplines in the critical task of achieving numeracy;
- and an awareness that chairing or heading a mathematics department takes savvy, acumen, high self-regard, and the ability to prove the Chain Rule.

A View from the Dean's Office

Cheryl Peters
*Academic Dean, Houston
Community College, Houston, TX*

What the administration expects from the mathematics department and, specifically, the department chair

Focus on student learning and student success. Excellent teaching is important, but student learning is the final measure. What additional support from the department do students need to move up and on? Keep student learning at the forefront of all your agendas.

Make sure that all curricula in the department are supported by a departmental syllabus that provides a structure for adjuncts and first-time teachers of a course. As a dean, I have seen perfectly dreadful syllabi handed to me when a student comes by with a faculty complaint. I have seen math syllabi and course outlines with numerous grammatical errors, little organization, no stated policies on exams, homework, lateness, absences, make-ups, etc. The chair should review all faculty syllabi on an annual basis to ensure that the institution is represented in a professional manner. The college or discipline should set guidelines or parameters for a good student syllabus, and, of course, the math discipline should determine the appropriate content and texts.

Give timely response to student complaints or concerns. This is a huge part of the math chair's job, but phone calls and emails must be responded to within 24–48 hours. Perhaps the problem can't be solved, but the student deserves the respect of communication that will give an idea of the process of solving a problem. Students are our core business, not a distracting annoyance. We can't ignore them, even if we can't always solve things to their satisfaction.

Do honest evaluations in your annual reviews of faculty. This is especially difficult when chairs are elected by their peers, but it must be done for two reasons: a) to prompt the ongoing professional development of faculty who are continually trying to improve their pedagogy and student success rates; b) to document and create a paper trail for those who stopped caring years ago about whether students succeed or not or whether they change one iota. Honest evaluations are crucial to the overall excellence of a department's teaching faculty. Show courage!!

Review data available from your Institutional Research office that helps you know your students, their opinions, their majors, the student evaluations of instruction, faculty grade distribution, etc. The more you know about the available data, the more informed you will be in making decisions affecting the faculty, curriculum, and students in your department.

Build schedules that are student-friendly, offering them flexible time formats and delivery modes. The old days of 8–5 instruction are gone forever; students want night and

27

weekend classes, web-based courses, web-tutoring, fast-track, self-paced formats. Build schedules that meet the needs of today's students, not an aging faculty.

Be visible in activities outside the department. The math department should not be the invisible department on campus. Encourage some faculty, if not you yourself, to become active in the Faculty Senate, a Curriculum Committee, or some other visible committee, so the math department can maintain a presence and voice on campus.

Sponsor a math club; sponsor math competitions; give internal awards or recognition to students doing outstanding work. Consider a yearly adjunct award if you are adjunct dependent. Whenever possible, engage in outreach activities at local high schools. Consider every activity a marketing/outreach possibility for your department.

Build relationships with faculty at high schools, two-year, and other four- year institutions. Stay abreast of the issues: course articulation, student retention, reform movements, teacher professional development, etc. Be a part of the larger math community in your area rather than taking an elitist or ivory-tower view of the world. You now represent your institution rather than your own parochial interests, and your institution needs to be a player in the local community. Also, these relationships will allow you to write proposals for large grants that require partnerships with several educational entities.

Let the dean know what the department's needs are. Don't suffer in silence. Do you need more full-time faculty, more staff support, more money to support tutoring labs, more technology dollars, another secretary? Communicate, communicate, communicate! The well-worn adage, the squeaky wheel gets the grease, is true. Become a tireless advocate for your core needs. The lesser ones will get prioritized with others in the college, but your basic needs should be forever at the tip of your tongue.

Always use the language of student success to argue for what you want. It is easier for the administration to ignore you personally but hard to turn its back on students who are struggling to pass mathematics requirements and who need additional support. Argue for what you want using the overall mission and goals of the institution, and generally these goals will mention student success or student learning.

Qualities you'll need to be a good/excellent department chair:

- Courage and backbone.
- The ability to see/take the larger view.
- Tough skin. Don't take things personally. You can't please everybody, but you can take stands that are ethical, equitable, and fair.
- Good communication skills, including listening, speaking and writing. Communicate horizontally and vertically. (And with your administrator, the rule is "No surprises." We can handle any bad thing, but we don't like to be surprised or blindsided.)
- Ability to build consensus.

How can the administration support the department?

The administration can help with money. Annual budgets typically prioritize institutional needs. If your department isn't on the radar screen, it will be business as usual for you. But if you have been an advocate for your department, money can flow your way. (For example, the HCCS Math Initiative: $250,000 provided for an additional hour in Math 0312, electronic classrooms, additional tutoring, stipends for curriculum revision, and additional faculty positions.)

The administration can help with department visibility. If your dean, provost, or VP is continually talking about student needs in the math department, that will increase your value overall in the institution. Remember to use the language of student learning and success. Also, let your dean know when student success is improving as the result of improved support. Success feeds on success; you won't find yourself cut off from resources, but perhaps even better supported.

The administration can help you with grants, new technology, improvements in your physical space, new hires, etc.

The administration will also support you in difficult personnel decisions and evaluations, especially if you have taken the time to document and keep your supervisor informed. It is very important to feel the support of your administration, and you should expect it if you are doing your job right. My first rule is to support the people I work with.

A View from the Dean's Office

Jimmy L. Solomon
Dean of Science and Technology
Georgia Southern University,
Statesboro, Georgia

So you are now the department chair. What is it that your faculty expects of you?

All faculty will expect you, at the very least,
- To be approachable and open;
- To be concerned about each one of them and their careers;
- To be consistent; and
- To be a good listener.

Remember that you are a member of the faculty, and that is where your primary loyalty lies. You must not lose sight of the fact that your jokes are not any funnier than they used to be, but you may get more chuckles from members of the department. But now you must be the principal fighter for the department. The faculty need to have confidence that you will win when pitted against other chairs at the table of the dean when resources are being distributed.

You are both a supervisor and a mentor. You have to be able to accurately assess the contributions of the members of the faculty and staff. There are times when you will address faculty in your unit who clearly do not carry the heavy load that others and you yourself do in working to achieve excellence for this department. You have to face that situation and deal with it. You might not be able to change it, but you do need to do a fair and honest assessment and to deal with it through review and merit rewards.

You are the principal fund-raiser for the department. You have to be resourceful, especially when state funding is insufficient for your needs. You should be able to secure funds from private contributions to help your department. As a member of the faculty, I was confident that the chair had a drawer of money that the chair could use if he/she so chose. After I became chair, I realized that it was the dean who had the drawer of money. Now as dean, I can assure you that the provost is the one who has the money.

These faculty expectations may seem rather lofty, but they are only one side of the coin.

Looking at the other side of your responsibilities, what is it that the dean expects of you?

Basically, as chair, you must convince the dean that your unit is more deserving of those scarce resources than the other units. Simply stated, you are the champion for the unit. Your success as an effective champion is greatly dependent upon your accurate assessment of the strengths and weaknesses of the faculty, staff, and programs, and

the ability to articulate those to both the faculty and the administration. Never hesitate to send your dean, provost, and president items that reflect upon the accomplishments of your faculty, staff, and students.

The academic department is key to any success in bringing about real change in the culture of the campus. You, as chair, are key to developing a departmental culture that provides your unit with the desire and the will to bring about changes. I offer the following thoughts that I like my chairs to think about.

- Develop a team vision for the changes—build real teamwork and articulate this vision.
- Develop a way in which you can successfully balance the individual faculty members' desires with the department's priorities.
- Develop clearly defined measures of productivity and rewards.

To be a successful chair, there are successful strategies that you can develop for representing your department and for reconciling the goals of the institution and the department. Be a team player, however, not to the point of not fully representing your unit. I tell my chairs that there is a line that represents conflict between the needs of the department and the college. I want them to get to the line. I do not have any difficulty in their sometimes going over the line—better that than stopping well short of this line. It is my duty to let them know when I believe that line is being crossed. Understand the mission of the institution, college, and department.

Every administrator must be consistent in his or her approach to issues. There are a couple of views as this relates to needs or requests. I recall as chair my dean saying something along the line of "Jimmy, if you ask for a dollar, then I know that I need to give you at least 95 cents, while if Chair X asks for a dollar and I give him a nickel, then he probably has 2 cents more than he really needed."

Work to foster stronger relationships with business, industry, and communities. Emphasize technology, especially in the area of instruction, and the impact upon courses and curricula.

Your efforts to secure resources will often compete with some units that will use the accreditation body argument. Note that the MLA (Modern Language Association) is not an accrediting body, but they surely help to keep small enrollments in composition classes. So I would carry a copy of the MAA's Guidelines for Departments and Programs to help make your arguments for the needs of the unit. In addition, I would advocate that you have a list of departments across the country that you would like to be compared with. You might consider two groups, a peer group and a group that you aspire to be more like.

Know your students. Nowadays, students are concerned about employment opportunities; they are career oriented; many hold jobs while attending school. In comparison to when we were in school, students take longer to complete a degree. They face increased college expenses that many are paying themselves, and more students receive financial aid. Recently I heard a keynote speaker make the statement that we do not need better quality schools, but rather more accessible ones.

Your efforts to be an effective chair will be inhibited by such things as:
- territoriality,
- faculty independence,
- heavy workloads,
- change which elicits fears,
- a feeling of no real authority, and
- tradition.

And your efforts will be enhanced by such things as:
- student demands,
- economic reality (once we were state supported, then state assisted, now we are state located),
- a clear sense of purpose,
- a clear focus upon the mission of the university, and
- a quality external advisory board.

In fact, it has been said that the majority of the negative forces come from within the department while the positive forces come from sources external to the unit.

Now for your interaction with the dean.

First you need to understand as best you can, the personality of the individual. Is he/she a morning or afternoon person? Is the individual data-driven or data-informed or something else? Does the individual want succinct or verbose arguments? How many times can you go back to try and change the dean's mind about an issue?

Know the issues in higher education. If you do you will understand where the dean and the administration are coming from. Some concerns which are currently on the dean's mind include:
- Restricted or reduced budgets,
- Numbers of majors,
- Distribution of credit hour production across departments/units,
- Increasing efforts to foster undergraduate research opportunities,

- Interdisciplinary efforts,
- Increasing grant activity, and
- Increasing emphasis on teaching.

It is really the case that the faculty sees this giant administrative serpent with you as its head; however, this unfair view is offset by the fact that the dean and central administration see you as the head of the giant faculty snake. If you are to be an effective chair, the truth is that you are both. I truly believe that the position of department chair is—with the possible exception of the president—the most difficult administrative position on campus.

However, I can state truly that my time as department chair was most enjoyable, and I hope that you will enjoy your time in this key role within your university.

Academic Administration and Liability

An Address by

Michael Anselmi
*University Counsel,
Towson University*

This talk identifies the administrative duties that involve a high risk of personal liability for college administrators, explains applicable legal principles, and suggests ways to reduce liability risks.

Liability

Who Can Be Sued?

An administrator can be sued individually for decisions made in the course of employment. If sued individually, an administrator is personally liable for any judgment obtained against the administrator in his/her individual capacity. Typically, private universities purchase liability insurance to cover administrators; state funded universities are usually self-insured. The insurance typically provides legal representation and if a judgment is obtained payment of the judgment. The university employer can be sued for decisions made by its academic administrators in the course of employment.

Administrative Functions that Involve Liability Risks

Review of recent litigation reveals that claims filed against administrators most often involve the following: faculty reappointment, promotion and tenure decisions, termination and/or discipline of employees, and resolving disputes among faculty.

Lawsuits arising out of the above circumstances typically involve claims of defamation; invasion of privacy; illegal discrimination on the basis of race, sex, disability, religion or national origin; or academic freedom/first amendment violations.

Defamation-Libel/Slander

Appointment and Tenure Process

The following cases involve defamation claims arising out of the appointment and tenure process.

Tacka v. Georgetown University, 193 F.Supp.2d 43.
As part of the tenure evaluation process, a department chair sought an external opinion on the scholarly work of a tenure candidate. The external reviewer concluded that substantial portions of the scholarly work were plagiarized. The department chair informed the tenure committee of the alleged plagiarism before the matter was considered by the Research Integrity Committee. Following a complete investigation by the Research Integrity Committee, the plagiarism charge was not substantiated. The faculty member, who ultimately received tenure, sued for defamation.

Other evidence in the case showed: (1) there was preexisting hostility between the external reviewer and the plaintiff; (2) the department chair selected an external reviewer who had a prior experience with plaintiff; (3) the allegation of plagiarism was made to the tenure review committee solely on the basis of the conclusion of the external reviewer; and (4) following an investigation by the appropriate committee, the plagiarism charge was found to be unsubstantiated. The Court found that the allegations of plagiarism was defamatory and that there was sufficient evidence for the issue to be decided by a jury.

Koerselman v. Rhynard, 875 S.W.2d 347 (Texas, 1994).
A department chair wrote a letter advising the dean that several female students complained to him about the tenure candidate's inappropriate comments to them. The department chair carefully documented the complaints and included them in a written evaluation summary that he placed in the plaintiff's tenure file. The chair also advised the plaintiff of the complaints and provided him an opportunity to respond to the allegations. The information was shared by the chair with the tenure committee. The tenure candidate was denied tenure and sued the department chair for defamation. The Court held that the statement was protected by a qualified privilege and granted summary judgment because: (1) there was documentation to show that female students did file such complaints and that the complaints were made directly to the department chair; (2) the information was relevant to the tenure evaluation; and, (3) the information was provided to a person (the dean) who had a legitimate interest in knowing about the complaints.

Larimore v. Blaylock, 528 S.E.2d 199 (Va. 2000).
A department chair and a dean individually wrote letters to the university's Board of Visitors urging that the plaintiff be denied tenure because of unethical publishing practices. The Board of Visitors had the final say on tenure. The plaintiff sued the dean and department chair for defamation. The plaintiff argued that the dean and department chair went outside the prescribed procedure by writing the Board of Visitors individually. Accordingly, he argued that the statements were not protected by qualified privilege. The Court granted summary judgment, holding: the statement about unethical publishing practices, even if defamatory, was protected by qualified privilege because the members of the Board of Visitors would ultimately vote on tenure. Accordingly, the individual members had a legitimate interest in knowing about the alleged unethical publishing practices. As long as the statement was made in good faith, it was not actionable.

Wynne v. Loyola University of Chicago, 741 N.E.2d 669 (Ill. 200).
The plaintiff was a candidate for departmental chair. A departmental faculty member wrote a letter to the dean concerning the plaintiff's "fitness" to serve as chair. The memo said the plaintiff often discussed her personal problems at work, particularly her "fertility problems" and "sleep disorder" for which she was undergoing psychiatric treatment. The memo also questioned the plaintiff's leadership ability and discussed the plaintiff's tendency to "wheedle, nag and domineer."

The defendant faxed the memo to the dean. All departmental faculty, however, had access to the "community" fax machine. No precautions were taken to protect the confidentiality of the memo. Another departmental faculty member, a nun, retrieved the faxed memo and discretely informed the dean by telephone. The dean requested that the faxed memo and a copy be delivered to him. A copy was made, however, another faculty member found the first page of the memo that had inadvertently been left in the copy machine and told the plaintiff about it.

The plaintiff who was not appointed chair sued the university and the faculty member for defamation and invasion of privacy. The Court granted summary judgment because the statements about the plaintiff's "fertility problems," her "sleep disorder" and her wont to talk about them at work were true. Not only did the plaintiff admit that she suffered from fertility problems and a sleep disorder, other faculty confirmed that she often talked about these matters at work. Truth is an absolute defense to defamation.

The comments in the memo questioning plaintiff's "leadership skills," and describing her as nagging and domineering were mere matters of opinion that could not be proved false. Defamation must consist of "factual allegations" that are capable of being proved false.

Because the plaintiff admitted that she often spoke about her fertility problems and her sleep disorder at work, she could not claim that she considered these otherwise private medical matters confidential. Accordingly, the disclosure of these matters to the dean did not "invade her privacy".

Would the disclosures about plaintiff's fertility problems and psychiatric disorder be defamatory if they were not true? Would the allegations that the plaintiff too freely discussed her personal medical problems at work be defamatory if not true? Had the plaintiff not herself disclosed her personal problems to her colleagues, would the disclosure to the dean be an invasion of privacy?

Points to Remember

The law recognizes that a meaningful evaluation of a tenure candidate—or any other candidate for academic appointment—requires the candid, forthright exchange of information without fear of being sued. The law protects administrators who disclose relevant information about the fitness of an academic appointee if: (1) the information reasonably relates to the fitness of the candidate for the position in question; (2) the information is properly documented or otherwise reliably supported; and, (3) the information is disclosed only to those persons who have a legitimate academic reason for knowing the information.

Liability Risks Related to Employment References: Negative References and False Positive References

Administrators are often asked to provide references for former employees. If a negative reference is given about an employee's fitness, there is a risk that the employee will sue for defamation. Positive references that are known to be false or that are patent misrepresentations, may, in certain circumstances, be actionable.

Qualified Privilege

The law recognizes that employment references given to a prospective employer, upon request, serve legitimate public and business interests; accordingly, a person will not be liable for providing such references upon request, if the reference is made in good faith.

Specific Cases

Hunt v. University of Minnesota, 465 N.W.2d 88 (Minn. 1999).
Negative statements indicating that an employee has "poor work attitude" and is not a "team player" were not defamatory. The statements were based on several years of working with the individual, they were primarily statements of opinion and they were stated in non-exaggerated language. Court held: "Employment references are conditionally privileged because the public interest is best served by encouraging accurate assessments of an employee's performance."

Bernofsky v. Administrators of the Tulane Education Fund, 2000 WL 422394 (E.D. La. 2000).
The department chair fired the plaintiff, a research professor. The plaintiff sued the university for discrimination. While litigation was pending, the plaintiff sought employment elsewhere and listed his department chair as a reference. The department chair did not provide the plaintiff any letter of reference. Because no letter of reference was provided, the plaintiff added a separate and additional cause of action to his lawsuit claiming that the University refused to provide him a reference in retaliation for his filing an EEOC complaint. Federal law prohibits such retaliation. The Court dismissed the complaint holding that there was no actionable retaliation because the plaintiff could not prove that the failure to provide a reference resulted in the plaintiff not getting the job.

Randi W. v. Muroc Joint Unified School District, et al., 929 P.2d 582 (S.Ct. Cal. 1997).
The defendant, a school administrator, wrote letters of reference on behalf of a teacher who had a number of complaints filed against him for "improper contacts with female students." The positive letter of reference said that the teacher had a "genuine concern for students," and concluded, "I wouldn't hesitate to recommend [the teacher] for any position." A female student was assaulted by the teacher after he was hired and she sued the person who provided the positive job reference. Appellate Court held that the defendant could be sued for negligent misrepresentation.

Points to Remember

Before providing any reference, require that the reference request be made to you in writing. Do not provide an oral reference under any circumstances. Any reference, including a positive one, must have a good faith basis. [If you provide an oral reference, your comments may be twisted or misquoted, and may be shared with the applicant and other interested parties. Without a written record, you may find yourself in a weak legal position should the applicant claim that your comments were incorrect and damaging. — Ed.]

Faculty Termination

Context

There are an increasing number of lawsuits filed by faculty alleging that adverse employment decisions were a result of a faculty member's exercise of free speech. In these cases, courts must balance a faculty member's rights to free speech (academic freedom) and privacy against an administrator's right to direct and manage an academic department.

Specific Cases

Collin v. Rector & Board of Visitors, University of Virginia, et al., 873 F.Supp 1008 (1995).

A faculty member sued the university and certain deans and faculty members alleging that he was denied tenure because of his outspoken "advocacy of minority recruitment" and his "opposition to race-related unlawful employment practices."

Court held that plaintiff's advocacy of minority issues constituted protected speech under the First Amendment. The Court further held that if plaintiff could prove that "but for" his advocacy of these issues, he would have been granted tenure, he could sue the dean and faculty members individually for violation of his First Amendment rights.

Miller v. Bunce III, 60 F.Supp.2d 620 (1999).

The plaintiff, a tenure track faculty member, received a $160,000 NIH grant to conduct research. Part of the grant money funded two other faculty salaries. The plaintiff complained to the department chair about the quality and quantity of the work done by the two other faculty members. The chair did not find any basis for the complaint. The plaintiff then complained to the NIH and reported that he could not verify the time charged to the grant by the two other faculty. NIH refused to pursue the matter. The plaintiff continued to complain causing considerable tension within the department.

While these incidents were occurring, the university changed its tenure procedures, which resulted in a delay in the plaintiff's consideration for tenure. The plaintiff alleged that the department chair and others retaliated against him because he filed a complaint with the NIH, thus compromising his tenure evaluation. Plaintiff sued the department chair and other administrators. He claimed that he had a first amendment right to complain to NIH and that it was illegal for the defendants to retaliate against him for properly exercising that right. Although the Court found that plaintiff's complaint to the NIH was "speech" protected by the first amendment, it dismissed the claim because the plaintiff could not prove he suffered an "adverse employment action."

If plaintiff is denied tenure can the plaintiff refile his complaint? Because only public universities are bound by the U.S. Constitution and thus the first amendment, do faculty who work for private institutions have any recourse in such situations?

Stebbings v. University of Chicago, 726 N.E.2d 1136 (Ill. 2000).

The plaintiff, a medical researcher at the University of Chicago, reported to the NIH that human subjects were inadvertently exposed to high levels of radiation. He alleged that he was fired in retaliation for insisting the university report the incident to the NIH. The Court dismissed the complaint. This decision was reversed on appeal. The appellate Court held that if the plaintiff could prove that he was fired because he reported the exposure of human subjects to radiation, then he was wrongfully discharged. The appellate Court thus concluded that a private university can be sued for the tort of wrongful discharge if it fires an employee for reporting university wrongdoing to protect the public.

Feldman v. Ho, 171 F.3d 494.

The plaintiff, an assistant professor of mathematics at Southern Illinois University, charged the chairman of the mathematics department with violating his freedom of speech. Specifically, the plaintiff, a tenure candidate, accused another faculty member of falsely claiming to have written a joint paper with a famous mathematician. The accused faculty member denied the charge. The department chair sided with the accused faculty member and the plaintiff was denied tenure as a result. A jury found for the plaintiff and awarded a judgment of $250,000 against the department chair and ordered payment of $185,000 in attorney fees.

The case was appealed. In a strongly worded opinion, the appellate Court reversed the jury's judgment. It held, without equivocation that the University's academic independence is protected by the Constitution just like a faculty member's own speech.

Quoting from a 1957 opinion of the U.S. Supreme Court, the appellate Court observed:

> [A university] must be permitted to determine for itself, on academic grounds who may teach, what may be taught, how it shall be taught and who may be admitted to study.

Relying on this premise the Court held:

> ... that for a university to function well, it must be able to decide which members of its faculty are productive scholars and which are not (or, worse, are distracting those who are). An unsubstantiated charge of academic misconduct not only squanders the time of other faculty members, but also reflects poorly on the judgement of the accuser. A university is entitled to decide for itself whether the charge is sound. Transferring that decision to the jury in the name of the first amendment would undermine the university's mission, not only by committing an academic decision to amateurs, but also by creating the possibility of substantial damages when jurors disagree with the faculty's resolution, a possibility that could discourage universities from acting to improve their faculty.

Wise department chairs make sure that they are well versed in the legal principles that apply to their responsibilities, educate the faculty appropriately about these principles, and act in good faith in all areas of responsibility. This does not guarantee that no legal action will ever be taken against the chair or the department, but it greatly reduces liability risks and provides the chair with valuable guidance in the handling of sensitive matters such as hiring and firing of faculty and staff, rank and tenure decisions, and student complaints.

Part 2

Case Studies

Introduction

These case studies are based upon real situations. However, you may feel that you know a particular situation because it is one that is fairly common. None of the names of institutions or persons are intended to be real; they are all fictional. If they are the name of a person you know or an institution, that is totally accidental.

These case studies have no answers or ideal solutions. Because they are thorny problems, they are interesting. You should think about what you would do or would have done to avert the situation. Don't take too much comfort in thinking this could never happen to you, however. Often a situation catches you completely by surprise and off-guard.

While reading these cases might give you some comfort, you can be sure that during your time as chair you will have, and probably already have had, situations that would make good case studies. Write them up. Doing so will help you think through how to handle a current or on-going situation or help you analyze what you did when you were dealing with the situation. If you participate in a workshop with other chairs, sharing your own cases will be particularly illuminating.

Students

The Non-placing Placement Test

The Disruptive Student

A Graduate Student Paddling Against the Current

TA Trouble

The Non-placing Placement Test

Mattingburg College is a two-year technical/community college with an open enrollment policy. The students are enrolled roughly 60% into technical programs and 40% into university transfer programs. All students who enroll in a certificate, diploma, or degree program are required to take a placement test. However, except for the nursing program, there are no mandatory placement cut-off scores. This policy is staunchly upheld at the direction of the highest levels of the administration.

Recommended placement cut-off scores have been established and widely published for all mathematics courses. Nonetheless full-time faculty members, who all act as academic advisors, use these scores only unofficially and informally when registering students for mathematics courses. As a matter of practice, about two out of three students who are enrolled in College Algebra are enrolled with authorization to override their placement scores. Consequently, the A, B, C success rate for this course is about 34%. The mathematics department chair feels that the corresponding D, W, F rate is detrimental to hundreds of students each semester.

Recently the chair worked with the college's Placement Committee to recommend a slight increase in the College Algebra placement score cutoff. When this report was sent forward, the Placement Committee was informed that there is no written policy in place for changing placement scores. So the department chair worked with the chair of the Placement Committee to develop a procedure for recommending changes to the placement scores for mathematics and English. Simultaneously the department chair sent a memorandum to the dean of instruction describing the desired slight change in the placement score for College Algebra.

The recommended procedure was returned with the comment that it could not be acted on this year. The chair's memorandum was returned with a note stating that until there was a procedure in place, his recommendation could not be considered. What else can this department chair do to insure that more students are properly placed in their mathematical studies? How can he better address the glaring case of massive misplacement of entry-level college students at his college?

The Disruptive Student

On the first day of classes for the spring term at Spruce University, the department chair approached Professor Larry Lunsford, a senior faculty member, informing him that there could be a potential issue with one of his students. He was told that the student was a problem in a previous class taught by Professor Stefan Young. Upon visiting the chair at his own initiative, Lunsford learned that the student had a history of disruptive behavior. Moreover, he may have taken and received an "A" in this same course repeatedly in the past. It appeared that all prior instructors had accommodated his every demand in order to avoid conflict. Lunsford, concerned upon hearing these new details, approached Young for more information and was told that the student typically engages the instructor in argumentative and/or incoherent diatribes on non-content related topics and makes other students feel uncomfortable and even threatened. Some of Young's students were willing to testify to this effect, though they had never been asked to do so. The chair had talked to the student and issued him a warning to behave or else be disenrolled from Young's class. Although the student continued to be a problem, he calmed down thereafter and eventually completed the course.

After the first diatribe in the classroom, Lunsford implemented a strategy to cut off the student's in-class comments shortly and politely. The student remained relatively quiet until after the first quiz was returned. After class he complained about the grading and not having had enough time; he demanded a grade change. Lunsford refused to change the grade, explaining that he expects students to be able to complete quizzes and exams in the allotted time, and suggested the student come by during office hours for help. The student indicated he would complain to the chair. Lunsford then wrote the student a formal letter with a detailed description of everything that had transpired in his class, sending copies to the chair and the dean.

At the end of the first exam, the student demanded extra time and refused to turn in his paper. At the end of the day, Lunsford found the student's completed exam in his mailbox. Upon getting back his exam with a failing grade, the student asked to be recognized and began to complain aloud about the instructor and his grading. Lunsford calmly advised the student to talk to him after class. When the student still refused to stop, the instructor told him that if he continued to disrupt the class, Lunsford would be forced to have campus security remove him from the class. Proclaiming that "it's a free country," the student challenged the instructor to call security. Lunsford made the call, and the student was removed.

The student was called to a meeting with the chair and the assistant dean for student affairs. At that meeting, the student restated his complaints and added, "You people have not yet learned anything from Columbine." Lunsford demanded that the student be denied access to his classroom, or better yet, be expelled from the institution. At this point, the campus police chief and the student judicial officer became involved.

At a closed meeting attended by all the academic players, multiple positions were articulated. The assistant dean wanted to make sure the student's rights were protected—the student should be allowed back in class "ASAP" with a warning. The dean said little, as if wishing the whole thing would go away, and stated he was willing to accept a consensus solution. The chair, after receiving numerous comments from other faculty members in the department in support of Lunsford, wanted the student suspended for a semester. The student judicial officer was torn between her sensitivity to student rights and her personal friendship with Lunsford. However, Lunsford remained firm in his demand. The campus police chief voiced the opinion that if something were to happen in that classroom after the student is allowed to return "someone will own this institution."

The university lawyer, concerned after hearing those remarks, suggested that the student be suspended for "three days to a week," and that the institution use that time to figure out what to do next. He also urged the chair to write a letter informing the student of this temporary suspension. The chair declined, and no one else would sign such a letter either. In the end, the student judicial officer was instructed to write the letter with the help of the chair and others, and submit it to the university lawyer for review. She met with the student and personally delivered the letter. The student threatened to sue claiming his right to free speech was being denied. What could have been done to prevent the situation from reaching this point?

A Graduate Student Paddling Against the Current

At Carderock University, new graduate students begin the program with varying levels of preparation. Some students come to the University because they have researched the areas of expertise of the mathematics faculty and find a match between their interests and those of the graduate faculty. Other students do not know what areas of research in mathematics they want to pursue and do not make choices until after taking coursework. Both types of students are accepted to the graduate program.

As REU (Research Experiences for Undergraduates) programs started to become more available, more students were entering graduate programs with some experience in doing research in mathematics. A new graduate student, Margaret Eagan, was one such student. However, her research was not in an area offered by the graduate faculty.

Eagan was very excited about the work she had done and wished to continue. She assumed the department would have faculty covering all areas of interest in mathematics and had not paid attention to the fact that the department did not have strength in the area of her research. Eagan was very disappointed to find the faculty she approached hesitant to commit to directing work in an area that they did not completely understand. She did not, in fact, find any member of the graduate faculty in the department who was willing to direct this line of research. But because of her family ties and personal financial situation, she had to stay in the area for her graduate work. In addition, starting her graduate studies somewhere else might cost her two years. The University seemed to her to be the only option.

How could the department help Eagan choose an area in which to work without totally losing the research work that she had already done and her enthusiasm for continuing in this area of study?

TA Trouble

Bonneville State University (BSU) has 45 tenure-track faculty and 75 graduate teaching assistants (GTAs) in the Department of Mathematics. Between 20 and 25 GTAs are new each year, and about 40% are international students. The percentage previously was higher, but after a fairly aggressive recruiting effort, the percentage of domestic students increased. BSU is an open admissions university and the department teaches many precalculus classes. Most GTAs run calculus recitation sections or teach their own precalculus class. The biggest complaint received about a GTA-taught classes is "My TA can't speak English." While occasionally justified, some faculty think that the real problem is that the students aren't trying very hard to learn or are unwilling to adjust to someone who is different and has an accent.

The mathematics department recently learned it would receive a VIGRE (Vertical Integration of Research and Education) grant starting next fall. Since VIGRE traineeships must be awarded to U.S. citizens or permanent residents, this new source of money would take some of the U.S. graduate students out of the classroom. Other universities were also trying harder to recruit U.S. students, and with so many institutions having a VIGRE award, it would be difficult to attract good U.S. graduate students unless they were given a VIGRE traineeship. How could these students be replaced in the classroom?

The department expected to need 30 new graduate students with about five of the new students receiving VIGRE support. Ironically, the percentage of new U.S. students might drop. In reviewing the applicant pool, it was clear that there was a very large number of international students, especially students from China, with superb mathematics credentials. The U.S. students who did not make first round awards appeared to be much weaker. Some faculty members thought that it was just that the U.S. students' backgrounds were weaker. If the department was willing to let them start in classes with the best senior undergraduates and allowed them to take longer to get their degree, they might prove to be just as good as the international students by the time they graduated.

There was an extra risk connected to making offers to international students. BSU required all GTAs for whom English was a second language to attend a three-week International Teaching Assistant (ITA) workshop that focused on spoken English and teaching skills. The workshop met the last three weeks before fall classes. To participate in the workshop a TA had to take and pass the Test of Spoken English (TSE). After the workshop, each student was required to teach a "mini-lesson" before a workshop panel. If the panel did not score the TA as "ready to teach", he/she couldn't be assigned any teaching duties.

Last year, five new GTAs did not pass the ITA panel assessment. The department had to support them for a full year and could only give them grading duties instead of teaching duties. Three improved their English significantly and are expected to pass the ITA panel this year. Two made very little progress.

The department is faced with a dilemma. Should it cut off support for the current GTAs who have not adequately learned English? How many new international graduate students should the department support, given the risk that they may not pass the ITA panel? Is it risky to increase the number of foreign students teaching classes? Should preference be given to U.S. students over international students who appear stronger mathematically? What is the right mix of domestic and international students?

To top things off, a member of the Board of Regents attacked the provost over the issue of foreign graduate teaching assistants. Apparently, one of the Regent's constituents told him that his daughter's math teacher couldn't speak English. An advanced graduate student who speaks with a strong accent taught the course. Despite his accent, he could be easily understood. The Regent is insisting that he be permitted to attend her class and hear for himself whether she has adequate English speaking ability. How should the chair respond to this situation?

Faculty: Dealing with Troublesome Individuals

Too Late to Change?
Socrates Too Far

Too Late to Change?

When Professor Suzanne Bixford first became department chair, she turned to old friends who had been leaders in the department, asking them to continue to take on the big jobs. She knew that Professor Jane Lyons had applied for the chair's position. She had not been on the search committee and had no part in the decision. All she knew was that she was the department's choice and Lyons was not. She asked Lyons to head an important job, which she did and worked hard on it. However, Bixford found Lyons very difficult to work with. She was always dissatisfied and constantly told Bixford that she was not getting enough from the administration, even though Bixford and others felt that the department had done exceptionally well. A former chair of the department told her that he and the dean had the same experiences with Lyons. She also learned that the dean would never want Lyons to be department chair. He asked her not to appoint Lyons to tasks that would bring her and the dean together. When salary decisions were made, she gave Lyons the highest merit raise in recognition of her current work.

After a year, Bixford realized that Lyons was no longer participating in department jobs or activities. She only taught classes and left. Bixford's secretary asked if Lyons was still in the department. When Bixford talked to her, Lyons said that she was reviewing her options and had a good offer to do something else. She told her that she understood, but that her merit increase for the year must reflect her activities that year. At the end of the year, Bixford gave her a low raise. Lyons protested to Bixford, the dean, and the provost. They backed Bixford, but the issue destroyed her relationship with Lyons.

A few years later, Bixford was going off for a year on a sabbatical and had to name an acting chair. She felt that there were unresolved issues with Lyons and could not consider her for the position. Lyons' behavior had not changed. She did little in the department, her teaching was uninspired, there were a few complaints from students each year, and her research was nonexistent. However, in an effort to re-energize Lyons, Bixford asked her to head an important department function while she was away. She explained that in order to become a department leader again, she must participate in professional organizations and must engage in scholarly activities including publication and presentation.

Lyons realized that she was not being asked to be acting department chair. She wanted to know why she was always passed over. She blamed the dean. When Bixford asked about her willingness to take the assignment offered, she responded that she didn't know that she wanted to do all the things this assignment would entail. Bixford explained that, if she were the chair, she would have had to accept responsibilities like these. Lyons claimed that, in that case, it would have been worth the effort. She expressed bitterness and told Bixford that she would have gotten the same recognition as she, and the invitation Bixford now has for her sabbatical, had she been chair.

How could Bixford mentor and advise Lyons, given her unrealistic expectations and her bitterness towards her? What should she tell her about her relationship with the dean?

Socrates Too Far

Every mathematics department has at least one idealist, the math purist who serves as the conscience of the department. Professor Emil Socrates is a senior member of the mathematics department at Hildenburg University. He is devoted to both teaching and research and is distinguished in each of the two areas. He strongly prefers, and makes his preference known, to work with students who are "serious" about their study of mathematics. For these students, he will devote enormous amounts of one-on-one time; his commitment extends even to the point of teaching as an overload, for no money or credit, an advanced undergraduate course or graduate course that does not have the required enrollment. He frequently oversees directed studies and student research.

Socrates likes to work with students one-on-one and is known to adopt a few students each year for special treatment. The university tries to encourage faculty to become more actively involved in working with students outside the classroom. There is a fund for faculty to have dinners and other social events with groups of students in their homes. Socrates finds that there are too many distractions at the office to have the concentrated time to spend with students doing serious study. Instead, he invites students to meet with him on a regular basis at home. He points to the philosophy of the university fund as a backing for his approach.

Occasionally, Socrates' interactions with students have raised eyebrows in the department. About eight years ago, he became very close to one of his students, a female of traditional student age. She was frequently in his office and it was widely known that they met together at his apartment. There was much talk in the department about their relationship. This student had taken some courses from other faculty members in the department, receiving average grades. Working with Socrates worked wonders, and she excelled in all classes taken with him and did fairly well in other mathematics classes as well. Some department members suspected Socrates of helping this student too much on her homework and independent work. The student switched her major to mathematics. Eventually the professor and the student were married. A few years later, they were divorced.

Now Socrates is mentoring another female student. The same scenario seems to be playing out. The student seems always to be in his office; they often work together at his home. This student failed freshman courses, more than once, until she took them from Socrates. Then she got A's. There is open talk in the department about the surmised relationship between Socrates and the student.

The college is located in a downtown area, with most students commuting. The average age of the student population is 27. There are no explicit rules against fraternization between students and faculty. In fact, there have been several marriages between faculty and former students or current students not in their classes. In this case, the student is currently taking courses from Socrates. What, if anything, should the department chair do?

Faculty: Part-Time and Temporary Faculty

Some Are More Equal

Keeping Adjuncts On Board

Adjunct Faculty—Welcome to the U.S.A.

Last Minute Adjunct Hire

Some Are More Equal

The University of Telanto has more than 17,000 students, and the mathematics department has about 30 full-time tenured or tenure-track faculty. In addition, the administration has allowed the department to hire several full-time lecturers. These lecturers are generally non- Ph.D.'s who teach 12-hour loads each semester. They have no responsibility for research, but they do committee work and participate in department meetings. They are eligible to serve on university and college committees. These are strictly temporary positions but may be renewed for a few years. They should not be retained long enough to have an expectation of long-term continuance of employment. The lecturers earn a full time salary, but are paid at the rate of adjuncts and get no benefits.

In addition to the lecturers, and because of the heavy service course load of the department, there are about 40 people who are regularly hired to fill part-time teaching positions during the day and evening sessions. Approximately 40% of the entire student-credit-hour load of the department is handled by part-time faculty. Over 50% of the course load is assigned to a combination of temporary lecturers and part-timers.

Several recurring problems present themselves with regard to these part-time faculty: proper hiring and dismissal; management and scheduling; orientation to both the university and the department; understanding of department goals and culture; and office space and support services. For example, many courses require the use of calculators and/or computers in classroom and laboratory settings. The use of technology must be done in a meaningful way that does not appear to be an add-on but contributes to student understanding and proficiency. Many courses require the use of calculators and/or computers in classroom and laboratory settings. In addition, all of these people must be observed teaching at least once a year and perhaps more.

There are several obvious problems in this environment. The first problem is the management of such a large and disparate staff. What departmental structure might make this work best? Second, what is an effective way of making the lecturers and part-time faculty feel part of the department? How can they be made to feel able to contribute to the teaching and content of courses for which they are mostly responsible—primarily service courses? Third, how does the department maintain standards for good teaching, for covering the syllabus, for grading, and for assessment?

Keeping Adjuncts On Board

The Mathematics Department at Wonderling College has always relied heavily on adjunct faculty who usually teach about 50% of the entry-level courses in any given semester. These faculty come mainly from two distinct groups: local high school math teachers who teach evenings, and "highway flyers" who teach part-time at several colleges and universities.

The department has a director for adjunct faculty who makes sure part-time faculty get information in a timely manner and keeps in touch with them during the semester. He observes classes of new adjunct faculty, handles their course evaluations, and works with the department chair to mitigate issues between adjunct faculty and students.

In the last ten years, the department has been rethinking its entire teaching enterprise, both in content and pedagogy. Initially, changes occurred in courses taught by the full-time faculty. As goals crystallized, it became essential to revise the entry-level courses and bring the adjunct faculty on board. Some first steps were relatively easy, or so the faculty thought. A team of regular faculty rewrote course syllabi to require use of graphing calculators and bought calculators with accompanying overhead projection units for all instructors. They encouraged adjuncts to use them at their other teaching jobs as well. But they underestimated the instructors' learning curves. So they started offering voluntary calculator workshops and encouraging adjunct faculty to drop in to a regular faculty member's office. This saved that initiative and provided the faculty with a much better understanding of the extent of support they needed to provide as they continued to make changes.

The faculty team targeted three introductory courses for a thorough revamping of content and pedagogy. Traditionally all sections of these courses were taught by adjuncts. In the revision phase, three regular faculty agreed to teach sections of these courses. Three adjunct faculty were invited to participate in the pilots. This required them to attend a pre-semester workshop and weekly debriefing sessions. Adjuncts who teach high school were specifically chosen. These teachers were under pressure to improve the mathematical competency of their students, and they viewed the workshops and debriefing sessions as professional development. Given the successful and voluntary participation of these three adjuncts, the department made the workshops and at least half of the debriefing sessions mandatory for all instructors of those courses.

The adjuncts who teach at several colleges were, as a group, the most qualified. Some taught calculus, discrete mathematics, algebra, and trigonometry. The department's goals and the university's general education goals for courses at that level included development of written and oral communication of mathematical concepts. For very good reasons, adjuncts that work at several places were most resistant. Some chose not to teach at Wonderling at all because of the time they would be required to spend in development workshops and debriefing sessions.

The university mandated that departments must provide an appropriate support structure for all adjunct faculty. What program would satisfy the mandate, serve to develop adjunct faculty, yet not be so onerous as to drive away the highly qualified and experienced adjuncts?

Adjunct Faculty—
Welcome to the USA

At Centron College, which is located in the suburbs of a major metropolitan area, hiring adjunct faculty is very important. There are excellent adjuncts who are either high school teachers with master's degrees or people in business and industry with masters or doctoral degrees. Most teach for many years and keep up with curriculum changes. The part-time faculty coordinator meets regularly with the adjuncts, individually and in groups. He designed a part-time faculty handbook, and conducts short workshops on major new curriculum and teaching approaches.

A member of the department asked that a newly arrived immigrant, Piotr Borisnykov, be given an opportunity to teach. Borisnykov had a Ph.D. in mathematics and had taught at a technological university in Russia for about ten years. The department chair was assured that Borisnykov's English was good, and, at the hiring interview, his language skills seemed adequate. He was hired to teach two courses, one in pre-calculus and the other a first-semester calculus course. It wasn't long before student complaints started, and then students began withdrawing from the courses. The chair referred the problem to the part-time faculty coordinator who noted serious problems with Borisnykov's teaching. The department chair also observed classes and found that the man understood limited English and actually taught very little, possibly because of the lack of command of the language. In a two-and-a-half-hour evening class, he drew one picture that he kept retracing. He said almost nothing, at least nothing intelligible. He did not seem to understand student questions. Students were patient and seemed to be trying to understand, but there was a mixture of boredom and frustration in their demeanor.

The department chair started working with Borisnykov, getting him to write and hand out notes on his lectures in the hope that this would help with his communications difficulties. However, his notes were almost as sketchy as his lectures. The department chair and part-time faculty coordinator worked with him to develop more thorough notes. More students withdrew. More students complained. Borisnykov did not seem to understand that it was his obligation to try different approaches and more explanation to get students to understand. About 90% of the way through the semester, the few students remaining in the class threatened to sue. The chair felt they were justified, seeing no way to salvage these classes at this late date.

What could be done to placate these students? What could be done to recover the lost semester in their mathematics sequence? What would be fair for the students who withdrew?

Last Minute Adjunct Hire

Part I

It was two days before classes were to start at Tennysohn Community College. Professor Janice Ponton, the mathematics department chair, was pleased that she had faculty to cover all of the mathematics courses to be offered by the department that semester. She was glad that all that had been settled weeks ago, so she could get on with the many other issues surrounding the beginning of a term. Then, one of the adjunct faculty contacted her to inform her that he would be unable to teach the course assigned to him. Ponton quickly contacted all of her regular adjuncts who might have been able to teach this course, but no one was available.

There was one more person on her list, Jacob Olstransky. He had applied to teach as an adjunct, but she hadn't had time to check out his references. So she placed a quick phone call to his main mathematics reference, someone she knew slightly at a neighboring university. Olstransky got only a lukewarm recommendation: "He taught a remedial course for us. He was competent, but students complained that he was sometimes impatient with them. I would hire him again, but he's not on the top of my list." The course that Ponton needed to staff was not remedial. It was an entry-level course in college algebra which used Excel to do various things with data. The course was fully enrolled, so Ponton didn't want to cancel the section

Ponton checked Olstransky's unofficial copies of his undergraduate and graduate transcripts. His undergraduate major was in secondary mathematics education. His coursework in mathematics was equivalent in number of hours to an undergraduate degree in mathematics. His master's degree was in middle school education with a concentration in mathematics. None of the graduate courses in this concentration were taken in the mathematics department. They were all specially designed courses on teaching and learning middle school mathematics. With a stretch in interpretation, Olstransky had the requisite graduate hours in the discipline to meet accreditation standards. Ponton felt he would make up in knowledge of good teaching what he lacked in mathematical background. Besides, the course he was to teach was hardly above the level of middle school mathematics.

Should she hire Olstransky? Or should she find some other way to staff the class? Or should she cancel it?

Part II

Ponton hired Olstransky. After a couple of weeks, it became apparent that he knew a lot less mathematics than Ponton had been led to believe. He didn't even know function notation, despite having the equivalent of an undergraduate mathematics major. As time went on, Ponton began to suspect that Olstransky had a medical problem. She would explain several things to him, but by the next time he had completely forgotten them. Students started complaining about a variety of things: he loses their homework; he marks things wrong that aren't; he insults some students in class. By the fourth week of the semester, Ponton knew she had to take action. What should she do?

Faculty: Personnel Action

The Contrary Mentors

At the Line of Battle

The Hazards of Multiple Reviews

A Hiring Paradox

No Place for Sara

I Have a Dream

The Contrary Mentors

The Mathematics Department at Montmercy College has 28 full time tenured or tenure-track faculty—about one-third in mathematical education, one-third in applied mathematics, one-third in pure mathematics. In the past, new tenure-track faculty members were assigned a three-person (all tenured faculty) Resource and Evaluation Committee. The role of the committee was to help the new faculty member get adjusted to the university and the department and to evaluate the faculty member in the probationary years. Members of the committee were expected to visit the probationary faculty member's classes, review his or her annual reports, and regularly meet to discuss progress toward promotion and tenure. The committee was expected to make a recommendation to the department's Tenure Committee regarding the faculty member's reappointment and eventual advancement to promotion and tenure.

At least one of the R&E Committees was not functioning very well. Members of Professor Alfred Graham's R&E committee had taken an adversarial role with Graham. Graham complained to the department chair, Professor John Stanwick, that the committee was harassing him and was not fair and objective in their assessment of his work.

In Stanwick's view, Graham's work was above average. His assumption of responsibility and his level of performance in handling new curricula and grants and his participation in leadership classes were seen as unusual for a new faculty member. Stanwick agreed with Graham that the committee was being narrow-minded and unfair.

How should Stanwick handle Graham's dissatisfaction with his R&E Committee? If he does nothing, the committee's harassment might drive a very good probationary faculty member away. Should he replace the troublesome members of the committee with faculty who might be more favorably inclined toward Graham? This might be seen as biased against those being replaced and would undoubtedly cause dissension in the department. What other courses of action are open to Stanwick in resolving this situation?

Stanwick faced another larger problem with the R&E Committees. With six openings in the department and four untenured faculty already in the department, the total number of untenured, but tenurable, faculty could be as high as ten in the fall.

How might the chair establish new ways of mentoring and evaluating so many probationary faculty, given limited resources and the department's strong aversion to change? How could the R&E Committees be structured or instructed to insure that they act equitably?

At the Line of Battle

Dr. Pete Reeves, a new young faculty member on an initial three-year, tenure-track appointment at Harewood University had done some teaching previously at a much smaller school. Reports of that experience, while not glowing, were positive. He seemed to be serious about teaching, and he seemed to be a likeable person. As with all new faculty members, he was assigned a teaching mentor. The mentor talked with him about his teaching. All seemed well.

Soon after the semester began, the department chair started to get complaints, but they did not seem to be too serious. Instead they were more of the form, "He didn't tell me we had to do it in a particular way, and now he is taking away credit if we don't do it his way." However, soon the complaints got more serious. "He is insulting us. You need to do something." So the chair asked Reeves' teaching mentor to visit the class. The mentor thought that the lectures were fine, but that Reeves seemed to be treating students like they were in elementary school. After meeting with Reeves and his mentor to discuss the problems and various strategies for improvement, the chair believed the situation would be corrected.

But the problems remained. One morning, as the chair arrived at his office, he was greeted by a delegation of irate students demanding something be done about this instructor. It seems that Reeves had asked students for feedback about the course and then had put the comments on the web, rebutting every one and naming the students who made the comments. From here, the situation deteriorated rapidly. Students started emailing parents with their complaints. Parents started calling the chair. One parent even threatened to take out TV and newspaper ads attacking the school and department unless Reeves apologized to his daughter. Reeves maintained that he was only doing his job, noting that the department had asked all instructors to survey students and to respond to complaints. He refused to apologize to anyone.

How can the department chair defuse this situation?

The Hazards of Multiple Reviews

At Upton College, faculty performance reviews are officially a mechanism for formative professional development. Since they occur in the spring shortly before merit raise recommendations must be made, they also provide the documentation that determines those recommendations.

The faculty performance reviews were never consciously coordinated with tenure and promotion evaluations. Until recently this was not an issue since the understood, but not well-documented, criteria were similar. Now the criteria for tenure and promotion have become more rigorous across campus, with the mathematical sciences faculty leading this change.

Two factions began to develop in the department. One group wanted the departmental criteria for faculty performance review to more closely match the criteria for tenure and promotion. They argued that cautionary feedback from tenure and promotion committees is watered down by good performance reviews based on softer criteria, and that the official faculty performance review, as the only mandated review process which affects tenured faculty, should have a summative component. The second group wanted the faculty performance reviews to remain formative. They argued that this was the only evaluation that recognized those faculty who commit themselves heavily to service, and the department and university in general would suffer if this were changed.

The chair liked that the current faculty performance review process allowed recognition and reward of faculty for their contributions even if those contributions would not carry much weight in tenure and promotions evaluations.

How can the chair lead the faculty to a resolution that retains the best features of the current process while minimizing the potential for mixed messages?

A Hiring Paradox

Professor Donita Rogers, Dean of the College of Arts and Sciences, was big on planning. She argued that the only way that Antalona State University would improve in quality was for each department to focus on a few areas where it had the capacity to build a strong group of research scholars. The department chair, Professor Rafael Gonzales, saw this coming and, three years ago, appointed a Hiring Directions Committee. The report of this committee strongly supported focusing all future hires around Combinatorics, Topology, Number Theory, and Applied Mathematics, areas within the Department which the report identified as "our areas of strength."

A separate issue concerned the number of women and minorities on the faculty. The state legislature passed a law that directed the university to make steady progress in hiring women and minorities. Benchmarks had been set and, if the university failed to meet them, it would lose 1% of its budget each year until it met the standards set by the state legislature. Many departments where there were more women and minorities (Sociology, English, Modern Languages, History) saw this as a way to get new faculty they would otherwise not be permitted to hire. They would identify a candidate and then propose the person be hired as a special "opportunity hire." It was a strategy that appeared to work but, as a consequence, it was more difficult for the dean to authorize hiring in areas (like mathematics) that needed faculty but in which there were few women and minority applicants.

The Department of Mathematics finally received authorization to hire one person, and a search was begun in Combinatorics. The short list included Tomas Richford, Ann Bowman, and Ralph North. Bowman and North were married to each other. All three candidates were outstanding. Richford had had a postdoc at UCLA and had published several strong papers, including one with a member of the Antalona faculty. Bowman expected to receive her degree at the end of the academic year. She was the best teacher in the group, and the advertisement listed teaching as an important credential. But her research was a bit far from the interests of the combinatorists on the faculty. North might have been the strongest researcher of them all, but he was a quiet person and did not have a particularly good interview. Gonzales sensed that all three would soon have lots of offers, and so he would probably get only one chance to hire any of them.

Gonzales also needed to think about the department's hiring priorities. Hiring two people in combinatorics was certain to play poorly with faculty in other priority areas who wanted a chance to hire in their area. One faculty member was heard to mutter "Two positions is an awfully high price to pay to hire a woman."

Faculty assessments of the candidates gave a narrow edge to Bowman, but the decision was Gonzales' as chair. Was it feasible to try to hire either Bowman or North? Should the offer go to Richford who had postdoc experience and was the closest to a sure thing? What about North whose letters implied that he would be a research star? Perhaps by hiring North, a case could be made for a second position for Bowman as an "opportunity hire." If the offer were made to Bowman, would it be more difficult to get a position for North? How might Gonzales reconcile the department to two hires in the same area? Indeed, should there have been any effort to hire one of Bowman and North, if the most likely scenario was the loss both of them and of Richford, too?

No Place for Sara

Elsbeth Sara was, in most peoples' minds, an excellent adjunct faculty member at Dimiter Community College. She was especially effective with the college's growing population of developmental students. Quiet, unassuming, patient and creative, she was able to help many students succeed in mathematics.

A dedicated full-time mother, Sara first came to the college as a part-time student. Her liberal arts undergraduate degree had included little mathematics. She had always enjoyed learning, and now that her children were in school, she decided she could devote some time to learning some mathematics. She excelled in all her courses and volunteered many hours in the math center helping other students. When she moved to a nearby university, she continued to work in the math center. Eventually she completed a second bachelor's degree, this time as a mathematics major, and went on to complete a master's degree, also in mathematics.

It was sometime during her master's program that the department chair, Professor Chad Wong, approached Sara with the idea of teaching a course. The college, like many community colleges, relied on adjunct faculty, and department chairs were always on the lookout for good part-time faculty. Sara's first class in pre-algebra went very well. It was the beginning of a long association. Over many semesters, Sara continued teaching while working on her graduate degree and after its completion. She became a dedicated teacher, open to new ideas and adjusting her teaching to meet student needs. She attended department meetings and participated in several summer professional development programs.

It would seem that Sara would have been a natural to be offered a full-time appointment, should one become available. She had many semesters of successful teaching at Dimiter. She knew the students, and the department knew her. It seemed a good fit.

An opening did occur, and Sara applied. But so did a lot of other people. When the hiring committee, consisting of most of the department, waded through the applications, many members were attracted to other applicants. In comparison to Sara, some had higher degrees and had studied more mathematics; others had more teaching experience. Some just wrote a better sounding application. They "looked better on paper," as one member said. The committee refused to consider Sara's many semesters of quality teaching with the department. The college decided that they really couldn't make a continuing hire that year after all, so Sara was asked, and agreed to take, a year-long temporary appointment.

When the next hiring cycle came about, Sara reapplied. Her application was stronger, but again the hiring committee did not consider her its first choice. Ultimately, someone else was offered the position, someone who turned out not to be very happy in the job, especially with the challenges of teaching developmental students and the open admissions policy. Sara decided not to return to part-time teaching. After only one year, the new hire was looking for another position.

Now the department is advertising the same position a third time. Wong knows that Sara is perfect for the job, but feels that the time has passed to offer her a position. Sara would know she was a consolation choice even if the department could be persuaded to go with the tried and true rather than be dazzled by paper stars. Is there anything Wong can do in order to hire Sara for the position?

I Have a Dream

Dr. Jane Schogren, a junior faculty member at Finehall University, approached the department chair, Professor Nelson Romberg, with an idea for an improvement in the curriculum in differential equations. She was enthusiastic and had some good ideas. She seemed energized by the prospect of working on the curriculum and was very concerned about student success. But she was naive about the time and effort that would be required to do what she planned.

There was always pressure from the administration at Finehall to improve curricula. Differential equations was a key area for majors and for service to the sciences and engineering. Thus, it was a very public course, receiving scrutiny and complaints from departments, faculty, and students in important service areas. When students in other areas complained, the complaints were likely to reach the dean. There was considerable dissatisfaction with the course, and no one else in the department had volunteered to look at the curriculum and make any real changes. Yet Romberg was aware of the likely response to ideas from such a junior faculty member. He was afraid that Schogren would end up working alone on the project and might become disillusioned at the poor response from colleagues. She could be setting herself up for derision and hostility rather than support and acclamation.

In addition, the reward system was skewed toward research. Only recently had curriculum improvement been considered important, but not nearly at the level of research output, especially for young faculty who must establish their research programs and reputation. Romberg knew the magnitude of the curriculum undertaking and suggested that this project might detract from Schogren's research agenda. Schogren assured him that she could handle the project and continue her research.

What steps should Romberg take to support Schogren in pursuing her curriculum ideas, and at the same time, make sure she is not neglecting her research and hurting her chances for tenure?

Curriculum

Jump-Starting Curriculum Change

Revising the Major

The Departmental Divide

Jump-Starting Curriculum Change

Professor Vladimir Solda was chair of a fairly traditional department at Holman College, a department that felt (fairly justifiably) proud of the job it did with its majors. On the other hand, the department was seen by the rest of the university as off in its own little world, unconcerned about its role in the larger picture of the university.

Because the department was fairly content, it hadn't paid much attention to discussions going on in the larger mathematical community about teaching issues, calculus reform, etc. However, Solda had, and he felt that the department needed to look at its curriculum, especially the lower-level courses taken by non-majors or by non-majors together with majors. If he were to do this in consultation with client disciplines, and modernize what the department is doing, there would be benefits, not only for the students but also for the department's reputation on campus, and, in the long run, for the department's general well-being in terms of support for facilities, staff, etc.

What can Solda do to get the department to buy into and participate in curriculum renewal?

Revising the Major

The Manford State University System mandated a review and revision of all undergraduate and graduate programs. In recent years, the university had moved increasingly towards job-oriented, applied programs. The largest majors were in business and education. Now that each program was to be completely revised, the administration made it clear that majors should follow the institutional goals to provide baccalaureate degrees that would prepare students to fill jobs in the state. This mandate did not preclude preparation for graduate study, but all departments were expected to offer a job-oriented alternative.

Many department members were not supportive of changing the traditional mathematics major. They felt it prepared a student well in mathematics and that mathematics is good preparation for jobs or graduate school. The department had tried offering an actuarial science course, but it was not well subscribed. They had tried tracks in a limited way, but those tracks followed lines of mathematical focus such as applied mathematics, algebra, or discrete structures. Department members felt that there were not enough mathematics majors to support a variety of tracks that were very different from one another. They noted that a job-oriented course had already proven to be undersubscribed by students.

The issue of revision of the major divided the department into three groups. Some members did not want to make any significant changes. Some were willing to consider modest change but did not want to pursue a radical change like the tracks approach again. Others saw an opportunity for complete overhaul and reconceptualization of the major. The last group of faculty consulted with a former mathematics department chair, now the dean. The dean was enthusiastic about collaborative majors and encouraged tracks, this time along interdisciplinary lines, each in conjunction with another department, particularly within the school. This third group told the other members of the department that they were working on their own recommendations in support of the dean's agenda. They declined to bring their ideas to the whole department because they felt they would not get a fair hearing.

Soon thereafter, the department faculty split into three warring camps, each distrustful of the others, and in several instances, not talking to the others. There was the group for radical overhaul, the group against significant change, and a group in the middle, some of whom tried to be peacemakers. Department meetings were very destructive, often with one group launching personal attacks against another. Meetings became a forum for the preservationists, who dominated the public conversations. The reformers, feeling increasingly threatened, retreated from department business. Their campaign was waged in offices behind closed doors.

The chair was told by the dean and provost to get his faculty to stop bickering and begin working together to accomplish the institutional mission to revise the major. They were under a deadline imposed by the system, which they had to meet.

How can the chair restore peace and trust and an atmosphere of collegiality so that the department members can again work as a team toward common goals that will appease the administration and be acceptable to a majority of the department faculty?

The Departmental Divide

Pines State University was converted from a four-year college to a comprehensive state university. Suddenly new expectations of faculty productivity were mandated. External funding had not previously been on the department's radar screen. Faculty teaching loads were high, enrollment was growing rapidly, and faculty time was consumed with committee work, especially search committees.

The first large venture into external funding was driven by the desire for curriculum change. College algebra and business calculus were the high enrollment courses, but were least popular among faculty and students. Revising college algebra was an ever-present topic among department faculty. Basically, this meant changing the order of topics and/or adopting a different text. The desire for change in both courses was aired at a department meeting with the result that Professor Donna Paige and Associate Professor William Oldham, both excellent teachers, agreed to head the effort to completely revise college algebra as a first step. The department chair, taking a page from the NCTM (National Council of Teachers of Mathematics) Standards, wanted a totally modern approach using technology, group work, writing, and real applications. The department gave its full support to development of a new course and agreed to try it on a pilot basis.

The project grew into innovative projects with funding from both NSF (National Science Foundation) and FIPSE (Fund for Improvement of Post-Secondary Education). The project completely absorbed the two Principal Investigators, Paige and Oldham. As time went on, with a few other department members actively participating, the team became isolated from the rest of the department. The revised course was tried experimentally. Student response was excellent, more students were successful, and the administration pointed with pride to this new approach and interest in teaching in the mathematics department. Meanwhile, Paige and Oldham published a text and traveled over the country giving workshops. They continued to get both NSF and FIPSE funding for curriculum development for other core courses and for continued improvements to the algebra course. Over an eight-year period they brought in over $1 million in grants. They were clearly stars.

What started out as department solidity degenerated into department disagreement. Paige and Oldham had little interaction with the department as a whole. They taught few courses because of assigned time on grants, and they expected the department to continue to teach the courses they developed. They often worked off campus, and did not get to know the new faculty who came into the department.

Some department members were dissatisfied with the new courses because they contained less drill and less traditional material. There was confusion as to what the department had agreed upon long ago, and some faculty wanted a new vote. The department chair, under whose supervision the project started, had taken a new position. The temporary department chair, unsure about setting policy, did not step into the fray. The administration wanted to continue the new courses and was pleased with the external support. The administration position was clear to the chair: they wanted to keep the new courses.

After two years, the temporary chair was named chair. By then, there was wide disagreement on direction. Department members were hardly talking to persons with whom they disagreed. Paige and Oldham were bringing in the majority of external funding in the department. Each had been recognized with university-wide awards in teaching and scholarship. Other faculty felt their own work was not being given appropriate recognition, especially those doing basic research and publishing in research journals. They claimed what they were doing is harder. The college-wide promotion and tenure committee required scholarly publications, so assistant and associate professors were confused by the different signals they were getting as to what was important.

The chair knew he must take a leadership position, but knew that any position would further alienate some department members. Dropping the new courses, developed with grant support and popular among students, would go against the provost. Research must be supported, especially for faculty promotions and tenure decisions. The researchers were frequently mentors to new faculty, and they were the most vocal opponents of the new curricula. Supporting both or neither group would surely deepen the departmental divide. What should the chair do to bring the department back together and find a mutually acceptable solution to the issues?

Leadership

Search Burnout

Coming In As Chair: Where Are Your Loyalties?

Departmental Management

Growing Pains

Staffing Summer Courses: Life in the Crossroads

Search Burnout

A subdiscipline within the mathematical sciences department at Snowdon State University was trying for the third consecutive year to fill a tenure-track faculty position. In the first two years, the candidates they wanted to hire were attractive to many universities and did not accept Snowdon's offers. This year the pool of applicants was small and only three applicants were chosen for an on-campus interview. After the interviews, the Search Advisory Committee decided unanimously that two of the finalists were unacceptable. The committee was split on the third finalist.

The senior faculty member who chaired the Search Advisory Committee argued that the third finalist had good teaching experience and, through mentoring, could become a productive faculty member within the department. A junior member of the committee argued that the success in both teaching and research of the faculty within this sub-discipline was based on teamwork. They held frequent meetings to keep each other updated on classroom issues. The group had several joint research efforts and regularly used each other to criticize individual research results. The junior faculty member believed the only remaining candidate would not adapt to this with any amount of mentoring and, being marginalized, would not attain tenure. But in the final vote, the rest of the committee chose to recommend hiring this candidate.

The disagreement was brought to the department chair, Professor Tyrone Williams, who agreed with the junior committee member's assessment. He asked the committee to reconsider its recommendation or present a mentoring plan along with a positive recommendation.

The issue that was presented to the dean was that Williams did not respect the faculty's role in hiring a colleague and, since he was not a member of the sub-discipline, he did not have the expertise to choose their colleague. The dean told Williams that, because the cost of searches was high, a third failed search might cause the department to lose the position. The dean recommended that Williams smooth the ruffled feathers and deal with the mentoring issues later. How should Williams proceed?

Coming In As Chair: Where Are Your Loyalties?

Marcus Oppenheimer was hired through a national search to chair a department at Newman University, an institution that prided itself on teaching. The job was particularly demanding because the administration gave the department many assignments, and these generally fell to the chair because the department faculty were unwilling to be bothered. The chair was expected to fulfill the faculty expectations in research, teaching, and service, with only a one-course reduction. There were some particular issues, which had been dragging on for some time, on which the administration insisted the department make progress. One of these, assessment, was an area of interest and expertise of Oppenheimer's. The department expected its new chair to take care of it. In fact, this was a major factor in their choice of Oppenheimer.

Oppenheimer made it clear to all that his primary concern was that students get a good education and that he would listen to all students who had a problem. He stated that any faculty member will occasionally have a disgruntled student or two and believed that the problem could usually be resolved with a little discussion or investigation.

In his interview with the provost, Oppenheimer was told that the biggest problem with the mathematics department was two faculty members who were widely viewed as the worst teachers on campus. He was not told which two, but he had a pretty good guess. They were senior members, tenured many years ago. When he met department faculty, they remained isolated, did not mingle, and were very negative and defensive when he talked to them one-on-one.

In the case of Paulo Kutz, one of the two faculty members, a trickle of complaints grew into a steady stream. About two-thirds of the way through the semester, one whole class, minus only one student, showed up en masse to speak to Oppenheimer. The complaints included Kutz's refusal to answer questions in class, telling students to come see him during office hours instead; assigning and collecting homework over material he hadn't yet covered in class; picking on a student who had self-identified as having learning disabilities; and not following the syllabus.

Some members of the department told Oppenheimer that Kutz only got tenure because, 30 years ago, the chair failed to send in a letter on time saying not to give him tenure, and the department has been living with this situation ever since. The second faculty member identified by the provost, whom Oppenheimer thought was merely a rather boring teacher, actively defended Kutz. He was worried that he was next in line if action was taken against Kutz. The rest of the department had learned to accept the two faculty and their on-going problems. They defended them on grounds of academic freedom. These faculty were fixtures in the department and had become symbols of rigor and high standards. In fact, their extraordinary view of "students as slackers" as the real problem was a cushion for the rest of the department faculty.

What should Oppenheimer do? If he does nothing, there will be four classes a semester from whom he will get regular complaints, and these students will leave Newman hating mathematics. But his department was watching him, trying to find out whether he will be a good advocate for the department. And, of course, the administration hopes he will be able to resolve this problem, which it had been hearing about for 30 years. Furthermore, how does he manage the job? How does he get the department faculty to work on the issues and share the load? Clearly, the issues are interconnected.

Departmental Management

Professor Allen Parkay was unanimously elected as the chair of the Mathematics Department at Altgeld College, a small, private liberal arts institution. He was a first-time chair and had only a general impression of what might be entailed.

Things had been relatively stable for the decade he had been in the department. There were no major changes in staffing, no major changes in the curriculum, and no disruptions in the pattern of who would teach what and when. The majority (roughly two-thirds) of course enrollments were in service courses, and the number of majors, while modest, had remained steady. Within the college community, members of the department were solid citizens. They served as advisers, supervised independent studies and internships, participated in admissions and outreach activities, and attended college-wide events. Each member of the department engaged in the profession and pursued professional development.

While the mathematics faculty felt that there has always been an institutional wariness toward mathematics, they were well-represented on major committees of the college and had reputations as good teachers. In fact, the departmental tradition of Wednesday afternoon tea with the mathematics majors was well-known and singled out as an example of faculty reaching out to students beyond the classroom. In addition, the department genially collaborated with the student liaison in celebrating holidays, organizing and hosting career nights, and contributing to liaison bake sales and other activities. Career nights typically brought a half-dozen mathematics alumni back to campus to speak. For many years the department organized and hosted an annual Mathematics Awareness Day for high school students.

On the other hand, mathematics enrollments had been falling over the last five years. This had been particularly noticeable in two service courses for professional programs that had been the bread-and-butter of the department. There had been a combined total of fifteen sections five years earlier; now there were only ten. One of the professional programs dropped mathematics as a required course in its major. The other program sustained a drop in enrollments as well as in number of majors—a casualty of a cyclical decline in the economy that was not expected to rebound for another three years. Also, the annual Mathematics Awareness Day had not really yielded new students.

The College had never had a formal peer review process for measuring teaching effectiveness, but members of the mathematics department had always felt free to discuss informally any aspect of their teaching with colleagues. There were monthly department meetings. Decisions were generally reached by consensus; the chair being viewed as first among equals. There was a predictable pattern of agenda items: identifying hiring and staffing needs, determining the teaching schedule for the coming year, writing individual annual reports for the departmental report and merit review, reviewing tenure candidates, developing the budget, determining the awardees for departmental prizes, and so forth. In many ways, the department seemed to be doing a good job.

As a new department chair, Parkay wanted to do a good job. He felt that the department didn't get its fair share of students, nor did it get the recognition it deserved, considering the contributions its members made, individually and collectively. The faculty was small and each member spent many hours on teaching, service, and professional development. No one was willing to consider or interested in making changes. They were happy with their professional lives. But the department could be severely cut if the current trends continued. No one seemed to know what to do, nor were they willing to take time away from all the other things they were doing to find out. They were not interested in making significant changes nor in making the effort that would be required. They did not believe such an effort would pay off. They hoped the problem would either go away or at least not affect them, since they were tenured.

What should Parkay do to retain the good work going on in the department, but also begin to resolve the problem of declining mathematics enrollments?

Growing Pains

In her six years as chair, Professor Mary Jenkins had led the department through turbulent times. When she assumed the position, the department was held in low esteem by the administration. It had accused the department of being elitist, out of touch, and not in support of the mission of the college.

Jenkins led the department in areas that brought it new life and energy and pleased the administration. Several tries were made to reform college algebra, pre-calculus, and calculus. Some of these were disasters, and some were successful. The most progress was made in calculus. With a reformed text, use of technology, and more attention to teaching, students were succeeding in greater numbers and complaining much less. Some members of the department did not approve of the changes, but all could see the economic benefits to the department. They got more positions, equipment money, and travel money than others; and because of large enrollments, the chair was able to stretch the budget to give more released time for research and special projects. In addition, department faculty had become quite good at getting grants, and there was a good bit of reassigned time to work on externally funded projects.

The department was much harder to manage now. The number of full-time faculty had more than doubled in six years. There was more research in different fields, meetings hosted by department members, and large grant-supported curriculum projects. All of these successes were taking a toll on the department. Faculty were doing their own things. Groups formed, and some faculty did not always approve of what other faculty in the department were doing, especially when it impacted them, such as curriculum reform. The department chair was busier, too. There were more faculty and students to manage, more complex issues, rise and fall in budget cycles, insistence of the administration on use of adjunct faculty, and faculty who were not performing. In addition, the chair had issues of assessment and peer and post-tenure review to figure out and implement.

Calculus reform became a major issue in the department. Faculty on each side of the issue wanted to bring it to a vote. Each side was lobbying hard. Clearly, a vote would only harden feelings of strife. Support for curriculum change was deteriorating; in fact, there was a backlash. Faculty did not see the improvements in student performance in later courses, as they expected. They were not willing to extend changes in expectations to other classes; thus a schism formed. If the department were to retreat from the reforms in calculus, the administration would likely withdraw extra funding to support equipment and smaller class size. Also, the reform-oriented faculty would be unhappy and feel defeated by those who "refuse to change." They felt the anti-reformists were undermining the good that had been done. They maintained the problem was that the non-reformers were steadfastly maintaining expectations from long ago in later courses. They claimed all the courses needed to be "fixed."

How can Jenkins adjust from a cohesive department of friends that was relatively easy to lead, to a large department that has lost the feeling of camaraderie and feels overwhelmed by directives from the administration? How can she provide leadership and direction at this critical juncture and diffuse the conflicts building within the department?

Staffing Summer Courses: Life in the Crossroads

Summer session was an important aspect of Crossroads Community College's mission to serve its community. Classes were popular with students from other institutions home on summer vacation and with newly graduated high school seniors seeking a head start on college. They were equally popular with faculty, some of whom were able to earn as much as 20% of their academic year salary through summer teaching.

For years when the department was small, the department chair, Professor Antoinette French, was able to make assignments so all who wished to teach could. Few classes needed to be staffed by adjuncts. Senior faculty were given some preference in selecting times and courses. In times when the maximum load was not possible for all, seniority also prevailed. However, there was a clear attempt on the part of French to balance loads and to be fair.

But with growth and financially difficult times came problems. New younger hires naturally wanted the opportunity for additional teaching in the summer. Because junior faculty earn lower salaries, the additional summer support was even more important to them. In the hopes of increasing revenue, the college added a second summer session and expected departments to offer courses in both. In another cost-saving measure, the administration decided to limit the proportion of course hours taught by full-time faculty. These changes introduced new levels of uncertainty for faculty and for the chair. There might not be enough classes for full-time faculty. Faculty might not be able to teach in the session or at the times they would like. More adjuncts were needed. It was impossible for French to guarantee at the beginning of summer that second session courses would obtain sufficient enrollment to assure that all faculty would be satisfied.

These new uncertainties brought out the worst in the department. While there were often hard disagreements about other department matters, faculty had usually been cooperative and at least polite. Not any more. Some faculty were willing to trust French with making assignments that would recognize student needs and administration restraints, and be as equitable as possible for all faculty. But others were not willing to do this. They insisted on new procedures, but were unable to offer any viable suggestions that took into account everyone's needs and desires. One faculty member suggested a lottery system with no memory from year to year. Another suggested that those who did not get the schedules they wanted to teach one year would have first pick the next year. (This particular person always claimed never to have received what he wanted and so always deserved to be first.) Still another maintained the right to be able to teach only at a specific time and not be "discriminated" against for this demand. Some senior faculty were adamantly opposed to not considering seniority first, as in all other matters. And on it went. No one seemed capable of compromise.

In the end, assignments were made—they had to be. Not everyone was happy, and they carried their grudges over into other areas. Attempts at rational discussion of summer school staffing during the academic year failed, with the result that the same arguments arose all over again in the following summer. How could French have avoided this repeat performance?

Campus Politics

Contract or Not to Contract

Dealing Across Departments

The Interdisciplinary Team

Dean vs. Chair

The Persistent Fund Raiser

Vision 21

Contract or Not to Contract

The hiring of non-tenure-track, permanent faculty (contract faculty) is a relatively new approach to staffing at the University of the Southland. In the last three years, the Administration has acknowledged that the reliance of the College of Arts and Sciences on part-time, temporary faculty at the freshman level works against the University's goal of improving retention. Tenure-track positions are not being created, but the option of adding contract faculty within the College of Arts and Sciences has been offered.

The Mathematics Department now has two contract faculty. Their obligations are teaching, service, and participation in professional development activities. Even with the best of planning, merging the obligations of the department and the interests and priorities of the permanent faculty with those of contract faculty would hit stumbling blocks. The following two issues have arisen during the first year:

- The definition of faculty in the Constitution of the University Senate is so narrow that contract faculty are excluded from university governance.
- Many important committees require faculty (as defined in the Constitution) membership from each department.

When the chair of the mathematics department raised concerns about the ability of contract faculty to fulfill their service responsibilities and to participate in university governance, the chair was told that there was no support among the permanent faculty of the university to amend the Constitution to include contract faculty.

Within the mathematics department, contract faculty had been accepted as professional colleagues, despite very grave concerns on the part of the department's permanent faculty that tenure-track retirement replacements had been frozen while the number of contract faculty on campus was growing. Within the department the first challenge to collegiality will occur in the fall when the department is required to staff committees. The contract faculty should, and will, feel like second-class citizens, and the permanent faculty will feel resentful that the contract faculty cannot take over some of these obligations. The solution to the constitutional issue must be found at the university level. How can the department chair keep the faculty focused on common goals so a schism doesn't occur?

Dealing Across Departments

A number of faculty members in the mathematics department at Faber University with an interest in mathematics education had been co-directing an NSF teacher collaborative grant for the last five years, focused on developing a new model curriculum for pre-service mathematics teachers. They designed two new methods courses that heavily incorporated discipline-specific content.

A new governor was elected with a platform pledged to improving the quality of public education in the state. His agenda consisted of new licensure requirements that included a discipline-specific, rather than an education, major for teacher certification. He also instructed the State Commission of Higher Education (SCHE) to review all baccalaureate education programs statewide, leading to an up/down recommendation for teacher certification accreditation. And he introduced new legislation that would expedite teacher certification for experienced professionals wishing to make a career change to teaching.

As a result of this newly mandated review, the College of Education submitted a university-wide program that called for all certification candidates to take a sequence of eight education courses, including two methods courses to be taught in the College of Education. However, the mathematics department proposed a revised mathematics major with an emphasis on secondary education. This major would consist of 41 hours in the major. In addition, it would require six of the eight education courses proposed by the College of Education plus the two new methods courses developed through the NSF collaborative and taught by mathematics education faculty. The Dean of the College of Arts and Sciences encouraged the mathematics department chair to compromise and to cooperate with the College of Education.

After a sequence of meetings with the provost, no compromise resolution emerged. Everyone found it objectionable to increase the number of required courses in the major beyond 41 hours, for such a program would be too demanding and likely very unpopular. It would force students to take more than 120 hours in order to graduate, an outcome almost certain to be unacceptable by the SCHE. The math department, under the leadership of the chair, believed the education methods courses were redundant and that the department's proposal was more directly aligned with the governor's initiative. They voted unanimously to take their case to the faculty senate.

The faculty senate decisively supported the mathematics department. But the provost declared the vote to be non-binding and refused to accept it. In the meantime, an editorial columnist with a conservative bent, writing for the leading newspaper in the state, published an article accusing the central administration of political correctness in supporting economically unviable programs that do not have large numbers of students but are popular among the "liberal faculty."

The department of mathematics has a relatively small number of majors. If the administration started looking for small programs for the chopping block, the mathematics major might be vulnerable. The College of Education Dean claims the faculty in education could not only handle the math methods courses, but would do it better than the mathematics faculty. Pulling these courses out of the department would be totally objectionable to the mathematics faculty who feel all reference to mathematical content will be lost. In addition, loss of these courses would further erode the enrollments in the mathematics department. Although the Senate backed the mathematics department, none of the other science departments teach their methods courses. The English department is the only other department in the university to do so. The mathematics faculty are seen by the provost and the education dean as holdouts and "exceptions." The provost, the respective deans, and the mathematics department chair have been summoned to attend a meeting with the president to discuss this whole matter.

How can the mathematics department chair convince the administration that mathematics and English occupy a role in the education of K–12 teachers that is different from that of other disciplines? How can the chair convince the administration that teacher education, in turn, is integral to these two disciplines, and that these methods courses should be the responsibility of these departments?

The Interdisciplinary Team

Reginald University has a rather substantial engineering program. In fact, most of the enrollment in first year calculus and physics courses comes from engineering students. After some discussion, Professors Danielle Straford, from mathematical sciences, and Philip Katon, from physics, began to work together to coordinate topics in calculus and physics. Their goal was an interdisciplinary course integrating single variable calculus with physics. Faculty from all three departments—Mathematics, Physics, and Engineering—were enthusiastic about the prospects from this collaboration. A pilot section was authorized and taught by the two professors. Students gave mixed reviews. Students who did well in both areas really enjoyed the integrated course. But even though the grades for calculus and physics were kept separate, doing poorly in one area made future scheduling difficult for the student. Thus, other faculty did not join the effort. Straford and Katon continued with the project and enough students took the option to ensure adequate enrollment in the paired sections.

The calculus and physics sequences spread over three semesters. Students liked the combined courses because they were assured of the scheduling for all three terms. However, in order to align topics, the topics in both calculus and physics did not follow the standard sequences. Thus it was very difficult for students to drop out of the combined sequence and continue in the stand-alone calculus and physics sequences. It was even more difficult for students to transfer into the combined courses from the traditional route.

After two years, Katon went on sabbatical. No other physics faculty member wanted to be involved in the combined math-physics sequence. Without consultation with the mathematics department or the dean, the physics chair cancelled the special physics section of the integrated course. The students who were finishing their sequences were forced to move into the stand-alone sequences. They were extremely unhappy and felt the university had reneged on the promise they would be able to continue through the whole sequence. They were out of sync in the stand-alone courses. Special help sessions were scheduled for them on missed topics. The experiment ended badly.

Now a few years later, individual faculty members are approaching the chair of the mathematical sciences department with proposals for a variety of interdisciplinary courses. As before, the initiatives involve one faculty member from each of two or three departments teaching aligned courses. How can the chair promote planning for interdisciplinary initiatives so as to provide for long-term support and continuation of the initiative independent of the individual faculty promoters?

Dean vs. Chair

After many years, the dean of Planchard College retired. This dean was very paternalistic and ruled with an iron fist. He circumvented reviews and tenure and promotion decisions, and found a way to sidestep every rule the faculty made to control him. Now the faculty felt they had a chance to make a real change. Professor Samuel Zabarosky, mathematics department chair, was appointed chair of the search committee. His good friend, Frank Carl, a member of another department and currently on assignment in the dean's office, was also appointed to the search committee.

Zabarosky had been hired as department chair, but also had faculty status and had earned tenure based upon his faculty position. At Planchard, chairs serve at the pleasure of the higher administration rather than for set terms.

The Committee and the College Faculty found a unanimous choice for dean. Dr. Blanche Greene came across as warm and very interested in the concerns of faculty. While she did not have administrative experience as the head of a university unit, she had served an administrative internship with the president of a similar institution. The president supported her hiring because of this internship experience about which she was very enthusiastic. In fact, the president often had interns connected with the national program in which Greene had participated.

As soon as the academic year started, things began to unravel. The dean began distrusting her closest advisors, believing that they were leaking information, and soon replaced Carl in his assignment in the dean's office. Zabarosky was furious and felt that he and Carl were purposely shut out of the inner circle. He felt that Greene distrusted him and saw his "edge" slip away. The dean asked her new assistant to report to her what people were saying behind her back. Greene and Zabarosky had heated arguments at the dean's meetings with department chairs. Faculty opinions of the new dean ranged from support to thinking the dean's behavior was bizarre.

Zabarosky had been asked by the previous dean to convert two temporary positions within his department into permanent, tenure-track positions. He had ignored these requests repeatedly. If the positions were to be turned into beginning tenure-track assistant professorships, neither temporary faculty currently in these positions would likely meet the qualifications. One was overqualified for a beginning position, having retired from elsewhere at the rank of professor. The other did not have a doctoral degree. The provost pressured the new dean to resolve the issue. Dean Greene got no further with Zabarosky than had the previous dean and made a unilateral decision to create two tenure-track, assistant professor level positions to be advertised for the next academic year. While the faculty temporarily occupying those positions could apply, Zabarosky knew neither would be hired because of the qualifications issues.

Zabarosky confronted Greene angrily and challenged her right to make such a decision. He warned the dean that her decision caused the elimination of the only two faculty of a particular religious minority in the department. Greene fired him as department chair on the spot. Zabarosky stormed off campus and refused to assume teaching assignments. Negotiations led by the provost with Zabarosky and his wife, an attorney, and Greene failed. Zabarosky sued on the basis of freedom of speech, religious discrimination, and unlawful removal from his position.

What were reasonable expectations on the part of Zabarosky in his relationship to the dean and his right to the job? What should each of Zabarosky and Greene have done with respect to the faculty positions, which Greene made into permanent positions?

The Persistent Fund Raiser

George Rodney, a senior faculty member and former chair of the mathematics department, had been a leader in securing external funding for the department through government grants and a regional network of corporate and community sponsors that he had cultivated over the years. He was highly respected and had broad support in the department, especially among those who had actively participated in his programs. He was particularly effective doing his own footwork, knocking on doors, and signing-off on handshake deals.

A new vice-president for the University's Institutional Advancement (IA) Office was recently appointed to replace a multi-year veteran who tolerated, and some would say, thrived in the "good-old-boy network" style of fund-raising. Following her appointment, the new vice-president introduced new policies, trying to change the culture and establish tighter management controls. One of these policies required that all grant requests and development activities be processed through her office.

Rodney had just received late approval of an NSF grant proposal, funded at a lower level than originally submitted. This grant would establish an educational collaborative with an inner-city school district involving its middle school students and teachers and the department's math education majors and faculty. It was the kind of outreach program the department had wanted for some time and an excellent "fit" with a new institutional strategic plan promoting similar initiatives. The grant would become operational in six months. Rodney was tentative about proceeding at the approved funding level but was ready to seek the additional monies needed to guarantee its viability.

The IA Office strongly advised the dean and the chair to ensure that institutional policy was followed in all fund-raising activities. The chair made it clear to Rodney that he had the support of the dean and chair, but that he must follow the rules. Rodney, feeling that he was being handcuffed by new IA rules, told the chair that he was considering withdrawing from the project. Other members of the department expressed concern about the extent to which their grants and corresponding support might be impacted as a result of IA interference. The chair convinced Rodney to try the new way.

Rodney acquiesced to a request that a representative from IA accompany him to a corporate meeting arranged by one of the professor's close friends. Following the meeting, the representative complained to the dean that Rodney had inappropriately detailed project budget information without prior approval. The chair met with Professor Rodney and asked him to work more closely with IA so as not to imperil his project and funding. The chair clearly wanted the department to appear to be supportive in its dealings with IA.

Rodney reluctantly agreed to pursue an IA lead and make a presentation following a given IA script at an open-house corporate function. He was surprised to find that a colleague from another academic department at the institution had independently scheduled a parallel presentation requesting funding for one of his projects. This happened without the prior knowledge of IA and in direct violation of one of IA's new policies not to approach the same prospective corporate sponsor for support of multiple institutional projects at the same time. However, IA appeared not to take any action concerning this infraction to its rules. Rodney was angry and again threatened to withdraw from the project and from fund-raising altogether

The department has a valuable resource in Rodney's fund-raising abilities, and the chair would rather not let that vanish. He is reluctant to complain too vigorously about the other department's behavior, since that department is an important client discipline that has threatened many times in the past to decrease their required number of core math courses. But still, he feels that he should do something about this perceived unfair treatment of Rodney. What should the chair do?

Vision 21

In his annual State of the University Address to the faculty of Estelle State University (ESU), President Randolf Jones challenged them to make it one of the top 30 public universities in the nation. He also appointed a task force to develop plans for what he called Vision 21. The Vision 21 task force met for over a year and then issued a report calling for new institutional focuses on graduate education and high quality research. They also identified a group of ten "target" universities and recommended that ESU strive to become more like them over the next 21 years. In general, faculty in the Department of Mathematics responded positively to the Vision 21 report. There was some concern that the focus on research and graduate education runs counter to the department's effort to embrace the recommendations in Towards Excellence and to make a greater commitment to the department's undergraduate program. Other faculty suggested that the department should put all its energy into research, even if that means shortchanging the department's teaching mission.

One byproduct of Vision 21 was a new campus push to rapidly increase external funding through an emphasis on interdisciplinary research and major centers. Another new initiative, begun by the provost, was called Quality Counts. That effort sought to make a list of honors, awards, journals and positions of leadership that were valued and could be used to measure the achievements of the faculty.

Especially controversial was the plan to identify the top journals in each discipline and then to count the number of articles published in those top journals by faculty or by graduate students. Each department was charged with identifying the top journals in its discipline. In an effort to provide guidance on this topic, the dean first suggested that each department identify "journals that would cause genuine celebration and perhaps special consideration at merit salary time when members of your faculty publish in those journals." When this caused even greater concern among the faculty, the dean signaled that she would accept up to 10% of the journals in any one field. In some disciplines, the Quality Counts initiative was viewed as a threat to academic freedom. In others, chairs indicated that they plan to submit a small list as evidence of their commitment to quality.

In mathematics, many faculty were engaged in making a list of outstanding articles that appeared in lesser journals. Many young faculty were concerned that when they are considered for tenure and promotion, the only publications that would "count" are those in journals on "the list." Other faculty feared that the lists would be used to compare different departments. Surely the fact that chemists and physicists publish more than mathematicians would do harm to the department when the counts for various departments are compared. Applied mathematics faculty were concerned that the entire effort will discourage interdisciplinary work—especially work that leads to publications in journals devoted to other disciplines.

A separate concern for the department was this new emphasis on external funding. About 40% of the faculty had some form of external funding, but most grants were small and the department's total annual external funding was far behind that of the lab science departments in the College. The department was clearly wrestling with the relationship between external funding and high quality research. Some argued that it is everyone's responsibility to seek external funds to support his or her research. Others said that their area is "out of favor" at NSF and that forcing them to seek funding would amount to forcing them to change the direction of their research to chase dollars. They argued that the university has a responsibility to support their research and that their publication record should be sufficient to document the quality of their research. Some faculty suggested that the way to respond to the University's emphasis on funding should be to put more effort into seeking funding for educational activities, because grants in education tend to be larger than traditional research grants in mathematics. Others believed that this would only detract from the need to increase research productivity.

A pending promotion decision appeared to have the potential to test how the department was responding to Vision 21, the challenge to focus on the pursuit of quality, and the administration's expectation that the department's external funding would increase. The case involved Associate Professor David Edwin, a tenured faculty member who was seeking promotion to professor and who had been in rank for eight years. He had a solid publication record comparable to that of faculty in the department who were promoted to professor within the last three years. Based on a report of the publication records of mathematics faculty at Big 10 Universities, his rate of publication was comparable to Associate Professors in the Big 10.

Edwin was also a solid teacher. Because he often taught large lectures, over the past decade he had taught far more students than the average faculty member. His peers respected his high standards, and his student evaluations might be described as good but not great. In addition, he was a good contributor to the service side of the department.

The one drawback, especially in light of the emphasis on grants, was that Edwin had not had a grant in over a decade. At a preliminary promotion and tenure meeting, some faculty argued that Edwin was a contributor to all phases of the department's mission and his achievements were comparable to those who came before him. A different view expressed was that Edwin' research program was workmanlike. He would never have research achievements that help achieve Vision 21 goals. Given his failure to obtain external funding, supporting his promotion would simply give the administration fuel for the argument that the department lacks a commitment to high quality.

The department (and the chair) clearly face a number of challenges:

1) How should they respond to the Quality Counts initiative, especially the need to identify a list of the top journals in mathematics?

2) How should the department respond to Vision 21 and the need for greater achievements in research and graduate education?

3) How much pressure should be put on the faculty to increase external funding and should this need influence decisions about hiring directions?

4) What is the right decision regarding Edwin's promotion review?

Legal Concerns

A Troublesome Dismissal

Free Speech vs. Discriminatory Behavior

A Troublesome Dismissal

Part I

A tenure track assistant professor, George Downs, in his third year was being considered for contract renewal. His publications and service were minimally adequate. However, his teaching evaluations had been consistently poor for three years as evidenced by student evaluations and peer observations.

Although Downs was generally liked by his faculty peers, his relationship with the department chair, Professor Tony Talbot, deteriorated significantly in the last year due to confidential discussions with him about his teaching deficiencies and his failure to improve. Also, within the last year, three female students separately complained to Talbot that Downs made inappropriate and unwelcome sexual advances to them. One of the students also filed a formal charge of sexual harassment against Downs that was being investigated at the very time he was being considered for contract renewal. Finally, a secretary in the office verbally complained to Talbot about the faculty member's sexist attitudes. That secretary had made similar comments in the past about male faculty that were unfounded.

Talbot discussed these charges with Downs and he adamantly denied them.

During the deliberations of the department's promotion and tenure committee, the department was split initially on renewal. Talbot, however, was not in favor of renewal. What if any comments should be made during deliberations about the:

- Verbal complaints made by the female students;
- The formal sexual harassment charge filed by the one student;
- The secretary's complaint about the sexist attitude?

If the department chair discloses the complaints and the disclosure tipped the scale against the faculty member leading to a non-renewal recommendation, then:

- What, if any, information should be documented in the promotion and tenure file about the complaints and the discussion of them during deliberation; and,
- Should the information be disclosed to other faculty and administrators considering the departmental recommendation?

Part II

The university accepted the departmental recommendation and did not renew Down's contract. After his termi-

nation, Downs wrote to Talbot complaining of his failure to properly mentor him and to support him during the promotion and tenure process. He concluded the letter by stating his belief that his contract was not renewed because of his unpopular religious beliefs and that he was putting the university on notice that he had filed an EEOC charge of discrimination based on his religion. Talbot had not been aware of any religious issues concerning Downs before this.

Shortly thereafter, Talbot received a call from an administrator at another university who happened to be a former colleague and friend. The administrator said that Downs had applied for a teaching position and was the leading candidate. The administrator asked for information about Downs' performance. After politely refusing to discuss the matter over the phone, the administrator indicated that without a reference from the mathematics department chair, Downs would likely not get the position and thus asked that Talbot provide a written reference. Should he provide the reference? If he chooses to write the reference what should he say?

Free Speech vs. Discriminatory Behavior

The department chair, Professor Murray McBay, was present during a hiring committee's discussion, and a member of the committee made a statement about a candidate. "Well, his age certainly is a factor: how can we hire this guy in a tenure-track position if he's likely to retire in a few years?" Is this illegal? What should McBay do about it? Should he remove the faculty member who made the statement from the hiring committee? If he leaves him on the committee, how can McBay be sure that, when the committee makes its recommendation, it hasn't engaged in illegal discrimination?

Part 3

Current Issues in Mathematical Sciences Departments

Introduction

The papers in this chapter were written by the discussion leaders of the June 2003 Department Chairs Workshop. The topics are issues of widespread concern, but not all will pertain to your department or institution. However, most will have interesting insights. There are many tips that are applicable not only to the topic of the paper, but to other situations as well. The writers were asked to frame their papers as if they were having a conversation with a colleague who is a new chair of a department of mathematical sciences. There are few assumptions made that the reader has been through all this before, and there is reflection on the topic of what the writer would want to share from his or her experiences. You should use these papers as a stimulus for your own thinking and for discussions with colleagues. They are grouped by the topics given to the writers. Again, there are no right answers or approaches, only suggestions.

Undergraduate Student Recruitment

Carl Cowen
Indiana University-Purdue University Indianapolis

Students

Undergraduate Student Recruitment
  Carl Cowen
  Bruce Ebanks

Graduate Student Enrollment
  Carl Cowen

By now it is a cliche that, for the good of the country, we need to be educating more mathematics, science, and engineering students. Moreover, since the traditional demographic for mathematics majors (white males) has been decreasing, we need to be attracting an increasing number of women and members of underrepresented minority groups to mathematics in order to stay even or improve on the numbers of mathematics majors graduating from our programs. But as a department chair, how much of your effort can you devote to things "that are good for the country"?

In fact, for your department to be successful, you need to be effective in undergraduate student recruitment. At the very least, the dean and the provost consider the number of majors when looking at sizes of departments, and usually they consider the number of majors as more important than the amount of service teaching. In addition, nearly every institution is intending to attract more members of underrepresented groups to the institution. If you can improve recruitment to your department, especially if you can successfully recruit larger numbers of students who are from underrepresented minority groups, your department will be noticed and (it is hoped) rewarded. Success in recruiting undergraduates to your programs is an important measure of the success of your department.

After succeeding in recruiting students to your program, you need to keep track of them to get an accurate count of your efforts. Don't trust the registrar to do this for you. Keep your own numbers and report them to the dean when you are asked for data. Of course, because the dean will be getting the registrar's numbers directly, you should also report those numbers and explain any discrepancies. For example, the biggest single group of mathematics majors in my department is the dual Computer Science/Math major. Because the registrar counts only one major per student, and because students must be declared CS majors to take CS major classes, none of these students are counted for mathematics.

Every request for data from my dean has included these students as a separate category in the total number of majors. Know the system at your college or university and work around it if you need to.

In truth, recruiting students to mathematics generally, and to your program in particular, is a very long and difficult process. For example, studies have shown that students begin making their career decisions (and therefore, choosing their college majors) in junior high school. Organizing a "girls in science" day where junior high girls come to campus to see the marvels of mathematics and science, or organizing a week-long summer mathematics and science camp for promising minority junior high or senior high students can be fun and rewarding. Moreover, participation in such events may be an important part of your majors' activities. At Morehead State in Kentucky, mathematics classes are cancelled one day each spring so that faculty and students can entertain dozens of visiting high school students and their teachers with exciting mathematics demonstrations, games, and contests. I'm sure the students feel good about their participation, and I'm sure the university and the department gets good public relations. But, on the other hand, these kinds of activities have a fairly low yield.

Without a doubt, your most important recruitment tool is retaining the majors you have and turning them into successful and satisfied graduates of your program. Deans and provosts are not impressed if you bring in lots of students as freshmen and most change to other majors as sophomores. Moreover, the word gets back: when Joe's older sister graduates from your program and gets a good job. Joe's friends, teachers, and counselors hear about it and are impressed. Many colleges get the first few days of Thanksgiving week as (formal or informal) vacation, but few high schools do. You can encourage your majors to visit their high schools during this week to tell the senior mathematics class or other groups of students about your wonderful program. Obviously, the success of such efforts is dependent on your success in working with the majors you have.

Clearly, the most important part of retention is the academic program for your majors and the success of your graduates in getting good placements in jobs or graduate programs. Students notice if their department is paying attention to them, both the faculty in their individual classes and the department as a whole. On a visit to another campus not too long ago, I talked with a junior mathematics major and expressed surprise that the department had not informed him of an opportunity he would have been interested in. Disgustedly, he indicated that they never tell the majors about anything. While he wasn't going to disappear from the program, if his report was accurate, the retention in the department surely would suffer. There are also less formal activities that can add some spirit to the cadre of majors. Perhaps your department can distribute MAA's Math Horizons magazine to all your majors, perhaps you can support an MAA student or Pi Mu Epsilon chapter, or perhaps you can invite successful alumni back to speak at career nights. Bucknell University's mathematics department sponsors "game night" once a week for all mathematics majors. Some of the games probably have some mathematics connections, but most do not, and the goal is for the students and the occasional faculty member to get together for some fun.

While I've never turned down a request to have a faculty member visit a high school class to talk about mathematics careers or our programs, it is not a high priority —the exposure is too small to be useful, in my opinion. It is much more efficient to spend time establishing strong connections with teachers in the schools from which your college attracts students or establishing contacts with alumni willing to talk with prospective students about your programs. If your faculty do visit schools, make sure they talk with the teachers and counselors as well as the students while they are there. A teacher and alumni network that you build will work for you year after year.

The critical group to be working with in recruitment is the group of strong applicants and accepted students during the period from November to February each year. These are prospective students who have indicated that they are serious about your program. Your yield rate for this group will be much higher than from more general blitzes. Studies by admissions officers have shown that students become increasingly committed in the late fall and early in the new year. By mid-spring, for most students, there is little you can do to change their minds. According to these studies, even in early December, a scholarship needs to be $2000 or more to grab the attention of the typical prospect who is leaning toward another institution. Phone calls by faculty, by you as chair, and by some of your more gregarious majors can be very important in bringing students from this group into your program. Many departments invite their majors to come to the departmental office for one evening a week during this period to call prospective students. They might be paid for their work, and certainly pizza and soda are in order. While prospective students will be pleased to receive a call from you or one of your faculty colleagues, they will most likely have few questions for you. But, they will have lots of questions for the students who call. On the other hand, there is a high probability that if you call, you will not reach the student but will reach a parent. Parents tend to have more questions for faculty and chairs than for students, and they will appreciate the interest in their daughter or son shown by the department.

Parents are especially important for you to win over. While students may not be pushed into what their parents favor, the parents can make it difficult for students to go into a field they oppose. If your college or university invites prospective students and their parents to campus in the late fall, this is an especially good time to make the arguments to the parents that majoring in mathematics is a good thing to do. Data about the successes of your majors in getting good jobs (and comparisons to other majors, most of which will be worse than yours) are especially important. Discussions of the ways in which your department pays attention to the majors are important. This is also a good time to bring up graduate school. Most parents don't know that going to graduate school in mathematics or science is not going to be a financial burden on them like law school or medical school would be. They are pleased to hear that departments pay students to go to graduate school. While it may seem premature, you may not have another opportunity to influence their thinking on this important point and, besides, they are flattered that you think their son or daughter might be graduate student material.

What about recruiting women and members of underrepresented minority groups? The same techniques apply. Make contacts with teachers and counselors in schools where you might recruit women and minority students. Have women and alumni who are members of underrepresented minority groups come for campus visits or make recruiting calls for the department. Make sure that women and minority students in your program have a good experience and succeed. If the numbers make it possible, organize social events or career nights for women and minority students majoring in mathematics, or perhaps in the sciences, in which the speakers can talk about the issues of studying or working in a field that is predominantly white and male.

Undergraduate recruitment is a critical activity for the department. You can really make a difference, and success in recruitment can pay off in important ways for the department.

Undergraduate Student Recruitment

Bruce Ebanks
Mississippi State University

One of the keys to successful recruitment of undergraduate students is getting them to visit the campus. A high school student who has set foot on your campus is, by virtue of that fact alone, much more likely to matriculate at your institution than one who has not. There are various ways to encourage students to visit your campus. One example is a "Discovery Day" for prospective students. Our Division of Enrollment Services organizes two of these each year—one in the fall and one in the spring. Prospective students are given a guided tour of campus and a chance to participate in various activities arranged for them. Basically the idea is to give them a small taste of campus life and to make them feel welcome.

Another way to get prospective students to visit campus is to organize contests and competitions with prizes for the winners. Some departments give small scholarships for the highest individual scorers in various competitions. Our university also awards course credit to students who score sufficiently high marks on final exams for several courses such as College Algebra, Trigonometry, and Calculus I. This year we gave a small scholarship to the student scoring the highest on the Calculus I exam. Next year we are planning to organize another competition with more of an outlet for creativity. As we learn more, we will continue to develop this plan based on our experiences. We hope to increase the number and amount of scholarship awards in the future.

Another effective means of recruiting is to design special events specifically for high school students who may have an interest in the discipline. For several years we have collaborated with a nearby institution in hosting a Sonya Kovalevsky Math Day for high school girls on our partner's campus. (This year we even allowed a few boys to come.) If you are willing to provide the organizational muscle, you may find support for such events from a variety of sources. This year we received full support from the Association for Women in Mathematics (AWM), but we had lined up support from a number of campus offices (Dean, Provost, Office of Research, Division of Enrollment Services) as back up in case the AWM support did not materialize.

This year for the first time we hosted a Sonya Kovalevsky Math Day on our own campus. As part of our advertising and recruitment for the event this year, we contacted high school principals around the region to issue invitations. Although we had not specifically targeted minority-rich institutions, we were pleasantly surprised to find that these schools responded to our invitation in relatively larger numbers. So in fact we had inadvertently instituted a means of recruiting underrepresented minority students. One principal at a predominantly minority school wanted to bring his whole school.

It is too early to know what long-range successes, if any, we might reap from this kind of activity. (As a side note, one minority student told me after last year's event that she was planning to attend our institution after graduation from high school.) We were certainly encouraged by the response this year in terms of attendance. In years past, our sister institution had hosted anywhere from 30 to 50 students at Sonya Kovalevsky Math Day, with the greatest attendance of 50 coming last year in 2002. This year (2003), for some reason, we had over 200 students! Part of the explanation may be that for the last two years we have held the event on Fridays, whereas in previous years it was held on Saturdays. It is certainly possible that students may be more interested in such an event if they can get away from their regular classrooms for a day, rather than giving up a Saturday. But it is also the case that we have scheduled a larger variety of presentations in the last two years, including some by faculty members from other departments talking about applications of mathematics in their disciplines. I would like to believe that students have responded positively to this effort, too.

Finally, there are more individually targeted outreach efforts. Our institution creates lists of high-ACT and National Merit Finalist/Semifinalist lists each year. Besides offering these students various university scholarship opportunities, the administration encourages departments in which these students indicate an interest in making contact with the students. I send each of those indicating an interest in mathematics a letter thanking them for their interest in the institution, providing information about the department and careers in mathematics, and asking them to contact me if they continue to be interested. These efforts have produced limited results up to now, but we did get one National Merit Finalist last year. She has turned out to be a real gem in our program. She has worked on a research project already in her first year, and she attended the regional MAA meeting to make a presentation. Even if only one such student is recruited, the effort may be worthwhile.

Some other universities have designed more extensive programs for underrepresented minority students. Uri Treisman has developed some very successful programs at the University of California-Berkeley and the University of Texas-Austin that have been copied and adapted by other institutions. They target highly qualified minority students based on standardized test scores, high school GPA, and leadership qualities. Treisman calls his initiative the "Emerging Scholars Program." The students are presented the opportunity to enter a challenging program that begins during the summer before they enroll at the university. One added ingredient that increases their likelihood of success is a social aspect. The students are placed into an environment where the study and discussion of mathematics is an integral part of their daily lives. This continues throughout their academic careers at the university. I cannot say much more since I have never been directly connected with such a program.

Graduate Student Enrollment

Carl Cowen
Indiana University-Purdue University Indianapolis

Just as recruiting is an issue for undergrads, it is also an issue for your graduate program. Indeed, if many of your classes are taught by graduate teaching assistants, it can be even more of a problem!

With the number of PhDs awarded to US students at about half of the total and the total number falling, we have reason to worry about future dependence on importing talent to fill the needs of our profession. Although the academic job market has been relatively unstable in the past decade or more, it has been my experience that US students with a PhD have been able to find jobs that come close to meeting their expectations. The job market has been excellent for PhDs in mathematics education, in statistics, and some other mathematical specialties, and very good for Master's degree students. My interpretation of this information, contrary to what I sometimes hear in the hallways, is that we should be encouraging our students to go to graduate school in the mathematical sciences—I see bright futures for students with graduate degrees!

There are many factors that are affecting graduate student enrollment, including the number of undergraduates majoring in mathematics. Some of these include the changing image of mathematics compared with other quantitative areas like engineering and computer science. As the leader of a department, you need to find your department's niche and exploit it fully, while being cognizant of the issues in the background over which you have little control.

Other things being equal, students will elect to go to higher ranking departments, choose fellowships or research assistantships over teaching assistantships, choose departments that offer areas they think they are interested in (based on their usually limited experience), and choose departments that are geographically compatible with their wishes. This is reasonable, and it is how I advise students! This forms part of the background for building your niche; unless you are Berkeley, you probably shouldn't try to fill Berkeley's niche.

NSF's VIGRE (Vertical Integration of Research and Education in the Mathematical Sciences) and the U.S. Department of Education's GAANN (Graduate Assistance in Areas of National Need) programs, though quite different, both provide traineeships for US students that allow departments to reduce the amount of teaching needed to support their graduate students. Without a doubt, these programs, especially VIGRE, have changed the mix of applicants to graduate programs—improving the number and quality of applicants to programs that can offer these traineeships and diminishing the number and quality of applicants to programs that cannot. If you intend to apply to NSF for a VIGRE grant or to the Department of Education for a GAANN grant, talk to people from departments who have them, get copies of successful proposals, and (for VIGRE) go to the meetings where representatives from NSF describe NSF's many programs, including VIGRE.

In most mathematics graduate programs I know, most students need to teach to earn their support. Since most PhD graduates and many Master's graduates go on to teach at some level, this is not altogether a bad thing. If you ask your graduate students to teach, you owe them the chance to develop their teaching skills. This usually means developing a program of workshops in which they can learn teaching skills and a mentoring program in which they can learn about their strengths and weaknesses and get advice on how to improve. It is desirable, but not always possible, to have them advance in their teaching responsibility, so that as they gain skill and experience, they have the chance to meet more challenges. Teaching will be good experience for almost all graduate students; our alums who work in industry come back and say that they benefited immensely from their teaching experience. Even though they say they have little formal teaching in their jobs, they do a great deal of informal teaching, like explaining to a management committee of the business school or liberal arts graduates why their technical project is important to the company and how it needs to be implemented to be successful. Having a successful program to help your graduate students learn to teach well will help them be comfortable while in your program and will make them more grateful for their graduate experience after they leave your program.

You may or may not be able to have a graduate program ranked in the top ten by the National Research Council (NRC), but you CAN have a successful graduate program in an appropriate niche! Success should mean educating the students you get to the highest level you can, within the constraints of their interests and ability, AND making them feel good about their achievement. To get a PhD requires intense desire and commitment as well as a dose of ability—not everyone who comes to your graduate program will end up having that desire. One of the biggest problems we have in our profession is that many faculty feel that it is not OK—but it IS OK—for students to want to get a Master's degree and then leave. We will all have more successful graduate programs when we feel, and pass the feeling onto our students, that a Master's degree is a fine terminal degree. In fact, it is an excellent degree in the job market. Master's degree students know a great deal more mathematics than Bachelor's degree students, and they aren't as scary to many employers as PhDs are.

One of the most important, but most difficult, parts of a successful graduate program is to help students be comfortable and happy with their achievements while challenging them to do their best in learning difficult material and tackling tough research problems. Setting standards high, but reachable, is the way to bring the students to your program and have them leave satisfied and recommending your program to others. This is especially true of women and members of under-represented minority groups. These students must feel accepted, encouraged, and successful for their younger colleagues to want to come to your program. Perhaps surprisingly, it is not just the faculty who must treat every student with respect—women and members of under-represented groups often feel left out of student life in a department and that sometimes leads to their leaving the program. The same can be a big problem between foreign and domestic students. Many foreign students come to graduate school with the equivalent of a Master's degree from their own country and have probably had a very intensive undergraduate program compared to most US undergraduate programs. In addition, they have different cultural expectations of what student life should be. Tensions can lead to an atmosphere where US students feel they cannot be good enough and drop out. Departments need to be proactive in preventing these differences from creating an unsupportive atmosphere in their graduate programs.

To recruit students to your graduate program, you would do well to establish good relations with faculty from other colleges that can be "feeder schools" for your program. Perhaps faculty in your department can regularly visit these schools to give recruitment talks to undergraduate mathematics clubs. In such a talk, it is important to mention that going to graduate school in mathematics or statistics does not involve a huge expense because they will be paid. Believe it or not, many undergraduates do not know this already. Of course, to establish feeder schools, their students must succeed in your program. Visiting one college with a large number of talented mathematics majors, when talking about bringing students to our graduate program, the department chair told me that they don't send their students to Universities X, Y, and Z because they just "chew them up and spit them out." Then he asked what would happen in our graduate program!

Nationally, we have a problem with recruiting to our graduate programs. In almost every college and university, we encourage our best students to go to graduate school. What about our second best students? Many of these students are good enough to benefit from graduate education, even if it is not education at Harvard or Berkeley or your alma mater. Our not encouraging them to set their goals higher deprives them of achievement that they would feel good about and might benefit them economically. Even more, it deprives our country of a huge resource of talented people. Changing this will be especially difficult because it happens in every class we teach, but trying to change will help us all.

Creating a successful graduate program is a very difficult task. It involves changing faculty and student attitudes as much as it involves changing rules and increasing recruiting activities. But it is a critical activity for the department in which you can really make a difference. Success in this arena will bring recognition to your department and make the department a better place. Perhaps even your dean will notice!

Curricula and Programs

Encouraging/Leading Curriculum Renewal
Norma Agras
Catherine Murphy
Martha Siegel

Undergraduate Major
Carl Cowen

Technology in the Classroom
Donna Beers
Michael Pearson

Program Review
Donna Beers

Encouraging/Leading Curriculum Renewal

Norma Agras
Miami-Dade Community College, Wolfson Campus

Curriculum renewal generally occurs naturally in a department that is composed of vibrant and progressive faculty. For the average mathematics department, however, the difficulties encountered while attempting to undertake such a renewal are many. Having led such a renewal in my own department, I can offer some advice in that aspect of the role of the department chair.

First, do not expect that your entire department is ready for a complete overhaul in terms of curriculum or teaching strategies. Generally, most departments have one or a few members who are more enthusiastic than others, more in tune with the latest trends in mathematics curriculum and methodologies around the country, and ready to embark on some changes. This small group is the beginning of the "critical mass," and it is the energy of these faculty members fueled by your support that will facilitate any change whatsoever.

Data can be helpful in convincing the non-believers that some change might be needed. Success rates, retention rates, numbers of majors in mathematics and mathematics-dependent fields can sometimes help people see the light in terms of the need for change. Consider presenting the data precisely at a moment when people are listening or at least expressing concern about negative trends in some of the areas cited above. You need to do this in such a way that many members of your department are in agreement that the situation might be in some need of improvement. Convincing a few people of such a need is the first victory in the battle for renewal.

Having faculty attend conferences is another helpful tool in this endeavor. Giving faculty the opportunity to get away from the every day life of the department, going somewhere away from home and meeting with other faculty in their field, is usually an invigorating experience. Faculty attending conferences can see colleagues from around the country present ideas and strategies with enthusiasm, and this enthusiasm can be very contagious. This type of opportunity is very critical for any change to occur. Further, faculty who travel to conferences should be given the opportunity upon their return to share with other department members their thoughts and experiences related to the travel. The faculty members, returning invigorated and with new ideas, need the opportunity to share and to try some of these ideas in your department. If such travel is deemed too costly, an alternative is to host a workshop facilitated by others experienced in some aspect of renewal.

Once enough faculty members are ready to make some changes, your support is essential. For example, you should make sure that faculty understand that trying new ideas will be looked at favorably in their performance reviews. They need to be reassured that, although student feedback is sometimes negative when faculty are trying new things, in no way will the faculty member's attempts to try out new ideas in the classroom have any sort of negative impact initially. You set the tone for innovation, and since you write the performance reviews, your faculty members need to know that their efforts will be positively cited in your evaluations of them.

The next body that needs convincing is the administrators to whom you report. If you are fortunate to have a dean, campus president, or provost who is a visionary, your efforts will be rewarded and doors will open for you. Otherwise, you need to present them with the data cited above and explain how it is crucial for the school or the college to embrace and support the renewal in your department.

You need to be clear about the needs that must be met in order for this renewal to be successful. For example, it is possible that certain rooms will need to be furnished differently (tables rather than desks, classrooms with at least one computer for every couple of students, for example). Retraining your staff support personnel in the new curriculum is vital for the necessary student support infrastructure to be in place. Toward this end, you might need to have meetings with your support staff letting them know of the upcoming changes. If you have tutors or teaching assistants, you might require that they attend classes taught using the new methods, textbooks, etc., so that they can be familiar with the material, topics, and methodologies and will better serve the students who are in those classes. It is important to have buy-in from the various facets of the department.

The faculty might need training, and you need to be sure that your institution provides that in a timely fashion during days and hours that are convenient to your faculty. Your entire faculty needs to feel comfortable teaching; otherwise, they will likely become resentful and resist and perhaps even do their best to sabotage your efforts.

The faculty who are involved in the initial implementation of the changed curriculum or methodologies need to feel that you support them even if things do not work well the first time. Discussions about failures are as important as discussions about successes if these are done without repercussion and in the light of looking for ways to make things work better the next time. Be open, flexible, and communicative.

If you are ever in a position to hire new faculty, look for faculty who have strong backgrounds in the field. These individuals tend to be more secure and self-confident and therefore are often less intimidated by changes. Further, look for faculty who seem to embrace change and new ideas.

Don't expect that you will have complete department buy-in. However, once you get a critical mass of maybe 40% to 50% instituting changes and the others see how things are working out, many of them will join the ranks of the initial agents of change in your department. Some faculty will always hesitate. Expect it and make those faculty feel as comfortable as possible so that, even if they are not active participants in the renewal, they do not get in the way of the others who are trying to progress to a more-up-to-date curriculum.

Encouraging/Leading Curriculum Renewal

Catherine M. Murphy
Purdue University, Calumet

Knowing the culture and current priorities of the institution and department is essential to successfully encouraging or leading change. This is especially true when the possibility of curricula change is raised. Faculty usually have a strong sense that what is taught, as well as how and why it is taught, is and should be their decision. Be aware that negative emotional reactions might occur.

It is also important to think about your own leadership style, both its strengths and weaknesses. I tend to be a consensus builder. I do not consider compromise a sign of weakness. I tell faculty and administration (in as light a tone of voice as possible) that since the sixth grade I've admired Henry Clay. Consensus building and leading the development of workable compromises takes time and energy. Sometimes externally imposed deadlines are not needed. At some point I have to put on my "boss hat" and drive the process. Even then, I keep in mind that I am faculty and want to be viewed as such by my colleagues.

A successful strategy I learned from the Chair of the Psychology Department is to write a stimulus paper about the major outcomes and steps to achieve them that I would like to occur and ask the faculty to critique it. I've found that if the issue isn't as important to the faculty as I thought it might be, the suggested changes are minimal. If it is important, the paper gets trashed, and they are so intent on telling me why I am wrong that most of their major concerns as well as lots of good ideas get put on the table without the usual interest groups attacking each other.

In 1998, after a few failed attempts to have the faculty read the national reports and think about what a contemporary mathematics major program at our university should look like, I tried the stimulus paper approach. I distributed my paper at our annual half-day retreat preceding the start of fall semester. I backed up my suggestions with information we had been gathering about how well our mathematics majors were meeting the outcome goals we had established in 1995. Not only weren't enough of the students meeting the goals, we weren't assessing them in ways consistent with the goals. I included a copy of the university's assessment guidelines as a reminder of the importance of addressing the discrepancies. Most of my curricular suggestions weren't radical, but I did recommend creating freshman/sophomore level seminars for the majors to help them develop a sense of identity as journeyman mathematicians. At our university, until they are juniors, mathematics majors are a very small minority in courses dominated by engineers. After some rather heated reactions and a lot of venting about everything, two senior faculty members volunteered to co-chair a curriculum writing committee. Several other faculty agreed to serve. I was invited to sit in as a non-voting member, which I did.

By the end of that academic year, proposals had been brought to the department, critiqued, revised, returned, and eventually approved. The faculty's sense of ownership encouraged them to suggest assessment strategies (and, hence, pedagogical approaches) for individual courses and the entire revised program. Because of the requirement of layers of approval for curriculum changes, it was the fall of 2000 before we implemented the revisions.

My recommendation for freshman/sophomore seminars for mathematics majors hit the cutting room floor immediately. However, a recent university mandate that each program must have a credited First Year Experience course motivated a faculty member to suggest that we go back and think about fleshing out freshman/sophomore level seminars as part of responding to this mandate. Development of this course, or courses, will occur in the fall of 2003. One of the best outcomes of this process is that no one now assumes that courses and programs will remain the same for decades.

Curriculum renewal involving other departments requires getting to know what they really expect from the mathematics courses that are part of their curricula. We faced that with our revision of the mathematics major because all engineers, science, and mathematics majors are together for four semesters. In general, faculty in these client departments couldn't articulate their expectations except as a list of topics. Since we needed, as well as wanted, their support at the School and University approval levels, I worked with two of my faculty members to summarize some relevant recommendations from national reports and gather a few different texts. The Chairs of Engineering, Chemistry and Physics were asked to distribute this information to their curriculum committees. We offered to visit one of their faculty meetings to discuss the proposed changes. Only Engineering invited us. However, all three departments supported our proposal.

What I've learned over the years is that successful curriculum renewal requires the Chair to do his or her homework—why we are thinking of changing, why this is the right time, etc.—before bringing up the topic. If one or more faculty members independently suggest changes, help them think through some of the challenging questions they'll have to answer. In instances where resistance is expected, try to have a timeframe that allows buy-in to develop. If other departments will be affected, work with them in parallel. Guide the process diplomatically.

Encouraging/Leading Curriculum Renewal

Martha Siegel
Towson University

Articulation and Recognition of the Problem

The department chair has to be the broker in the area of curriculum renewal. The first question to address: is there truly a need to change? Many department members prefer to leave well enough alone, as their teaching of elementary courses, especially service courses, is sometimes seen as a safe and routine part of their teaching schedule. Redesign of the major and redesign of upper-level courses require phase-in and depend on the willingness of the administration to allow for experimentation (and perhaps small classes) while the major is revamped. Beginning new graduate programs is another difficult and intensive project. In every case, the chair is responsible for the collection of relevant reading materials, background information, and data.

The problems with considering changes in elementary courses include the following:

- Several departments may be using the same course for their major requirements and yet have different needs when asked about desirable outcomes. It is important to get their input. Once you ask about what each one wants, you have an obligation to design courses to meet the stated needs.
- Graduate assistants and adjunct/part time faculty are frequently the primary teaching force for elementary courses. Doing anything innovative costs the department both time and effort to prepare these people to teach new courses effectively.
- Asking for faculty to dedicate time to such curriculum renewal may take a lot of persuasion since the rewards for redesign and renewal efforts are rarely recognized for promotion, tenure, and merit.

Renewal of the major is a large undertaking as well. Program review can help to generate interest in such renewal, but the prospect of an impending program review may also deter a department from trying something new and innovative that may not fare well in an upcoming review. Knowing how to approach the needs of all the students in the major—some wanting to teach, others going into industry or government, and others heading to graduate school—is a serious undertaking. A department should have clearly stated objectives for its major programs. An Advisory Board of people in the field can be very helpful in this endeavor. Departmental Advisory Boards can pave the way for reform of applied programs, but also many of those on the board may be very supportive of the basic real analysis, abstract algebra, topology, and geometry backbone or core of the major.

Another part of the design of the major is the development of one or more capstone courses. Should all students be doing research and at what level do we require it? We have an ever-present need to assess the effectiveness of the majors we design. Stating clear outcomes should help in the assessment process. At our institution, any new program must have an assessment plan. This should also be true for revisions in old programs.

When a department wants to initiate collaboration with another department, designs of interdisciplinary and cross-disciplinary programs present their own set of obstacles: how to initiate the work with other disciplines, and how to prepare faculty members to teach in interdisciplinary or cross-disciplinary courses. These are thorny issues that should be addressed up front. If not answered or addressed, they can create scenarios for failure. How can the administration help? How can the MAA help?

Structural Questions in Approaching Renewal

One approach is to appoint an ad hoc committee to consider the problem and suggest a curricular solution. Who should be on such a committee—senior members of the faculty? newly arrived assistant professors? students?

Encourage the committee to read available materials and consult with colleagues in other schools to see how the problem is handled elsewhere. Urge the committee to apply for funding outside the department for the purpose of studying and recommending reform. Have the committee present its work-in-progress to the department at regular intervals to get a sense of the department's reaction to suggested curriculum change and departmental guidance to those changes as they work, rather than reaction (which could be negative) to a finished product. Content and pedagogy should go hand-in-hand. Consult the client disciplines. Follow best practices. Evaluate the plan for clear objectives, the assessment component, and cost.

In any process of curriculum change, you must consider how to gain consensus for the plan; how to get department members on board; how to get client disciplines to endorse the plan; how to get administration support. What if the cost is significant? Such costs as undergraduate research requirements for all, small capstone courses, or added credit hours to courses that affect teaching loads can be substantial. Are new faculty needed? Can the department have a successful search before the program is in place? Will senior faculty accept the plan for faculty retraining?

Undergraduate Major

Carl Cowen
Indiana University-Purdue University Indianapolis

Why do we have a mathematics major? How do we want our students to change because of it? Given the population of students at our college, who do we want to major in our department? How many majors do we want? These are questions that your department must address as you consider the structure of your program for undergraduate majors.

Many comments related to recruiting undergraduate students are intimately related to the comments here—after all, students can only be recruited to a program they see as meeting their needs. If you have a successful program and you have reasonable publicity, you will have majors appropriate to what you accomplish in your program.

How many majors do you want? This is a serious and strategic question: deans and provosts usually pay attention to the number of majors in a department. At least part of your answer depends on what the consequences of the department's actions in this regard would be. As a part of full disclosure, let me be forthright about my prejudices: I believe more is better! The only exception to this rule would be if an increase in the number of majors would put undue stress on the department's resources that would not be met with an increase in support from the administration. In particular, I do NOT take the point of view advocated by many of my friends and colleagues that we want to have majors who are more or less like us. Hardly any of our students are "like us," so proposing that all our majors should be "like us" will lead to a small number of majors. Instead, I believe that a successful department will have many "weaker" students as majors, and I believe that this is OK because these students have much to gain by majoring in a mathematical science and our department will be judged primarily by the best students we have, not by the weakest.

With this point of view, we should expect the majority of our majors to enter the work place on graduation with a bachelor's degree, either as teachers or in jobs in business, industry, or government. A smaller number will plan for graduate or professional school, for example with the goal of a Master's degree in some area. An even smaller number will want to get a PhD in a mathematical science, "like us." With this perspective on where our students are going, our programs need to help them meet their goals and will therefore NOT be geared primarily toward preparation for getting a PhD in mathematics. I believe our department, our students, and our society will be better off if we educate more students in the mathematical sciences. First, I believe their mathematical training will be an asset to their careers whether their job title is business analyst, quality control engineer, systems analyst, teacher, or professor, and the evidence I've seen shows that mathematically-trained graduates do very well in the marketplace. Also, I believe that mathematically-trained teachers will build a stronger foundation for the next generation and that mathematically-trained workers in business or government will bring different perspectives to their jobs in contrast to students from business, liberal arts, or other science backgrounds.

It should go without saying, that in designing an undergraduate curriculum, your department should be aware of the professional reports on related issues, especially CBMS' (Conference Board of the Mathematical Sciences) report, The Mathematical Education of Teachers, the report of the Curriculum Foundations Project, and the forthcoming Committee on the Undergraduate Program in Mathematics' CUPM Curriculum Guide. (See the Bibliography in the Appendix.)

It should be clear by now that, whatever students' majors are, they should be familiar with the use of computers as tools in that area. We should be building the use of appropriate technology into the programs of our majors—our students will be more comfortable than our faculty in doing so!

In preparation of teachers, I'm a strong advocate of the point of view that they should know mathematics beyond the level they will teach, and that they should have a deep understanding of the mathematics that they do teach. In particular, I believe it is OUR job to make the connections explicitly in our classrooms, between more abstract mathematics and the algorithms and concrete mathematics they will teach. For example, they should be comfortable with the statement in the fourth grade arithmetic book that $1/6 = 4/24$ and my statement that $1/6$ and $4/24$ are not the same, and be able to tell a story about why they are and why they aren't. They should know why their high school algebra teacher insisted that $1/\sqrt{2}$ should instead be written as $\sqrt{2}/2$ and why it is an anachronism. Finally, they should have an adequate concept of the nature of mathematical knowledge, its experiential and its deductive sides. Our curricula for teachers should include these connecting ideas in addition to the "how to" courses and student teaching from our colleagues in the Department of Education and the "math major" courses in algebra and analysis that all our students take.

Students preparing for jobs in business, industry, and government need very broad backgrounds. Alumni and recruiters I've talked to value the modes of thought that the study of linear algebra, real analysis, and modern algebra brings, and they also value the tools that students learn in a statistics course, a discrete mathematics course, an optimization course, or a modeling course. It is a real challenge to create a program that is broad enough to include the basics of the major and an exposure to the mathematics that provide the important tools of thought in the modern world.

If your department includes statistics, you have the opportunity to encourage your majors to concentrate in statistics, which is an increasingly important, but often neglected, mathematical field. If statistics is in a separate department, programs leading to a double major should be constructed and promoted by each department.

It appears critical that mathematics departments begin cultivating double majors in other disciplines. Double majors in mathematics and computer science or mathematics and physics are already quite common. In addition, we should be thinking more broadly and welcome double majors like mathematics and biology, mathematics and chemistry, or mathematics and business (as in an actuarial program). Challenges in such programs include keeping the size manageable while including the breadth that each department wants and getting credit from the dean or provost for teaching these double majors.

As a profession, I believe we need to be encouraging more of our students to go to graduate school in a mathematical area. In addition to sending our best students to graduate school as we've been doing for years, we should encourage our second best students to consider Master's or PhD degrees and our merely "good" students to consider entering a Master's program. Not only will they learn a great deal more mathematics if they do so, but more of them will discover that they really should be going on for a PhD. If we have a sufficiently large group of students "like us," it makes sense to provide a major track for that group, but for most institutions, I believe our broad program should prepare a large number of students to pursue further courses at the master's level. If we do so, I'm confident we'll be rewarded by a larger number of students who eventually get excited enough to get a PhD.

Our undergraduate majors should be learning the facts of a large body of mathematics and the modes of thought that make it possible to solve mathematical problems. At whatever intellectual level our students enter, they should leave with a broader knowledge base, sharpened problem solving skills, and greater ability to think deeply about mathematical problems. We should measure the success of our programs, not in the absolute achievements of our students, but in the improvements our majors have made during their time in our department. It is this belief that lies at the heart of my desire to have a large number of majors including many who are not "like us." If we are creating outstanding undergraduate programs in the mathematical sciences, the future will be bright for students who leave these programs. With graduates with bright futures, it should be easier to bring more good students into our departments.

Technology in the Classroom: Winning Faculty Buy-In

Donna Beers
Simmons College

Researchers who are studying today's college students say that the digital world is changing the way they learn. Carole A. Barone, vice president of EDUCAUSE where she is responsible for the National Learning Infrastructure Initiative (NLII), writes: "Students 'Think Differently.' Students who take email, instant messaging, and seemingly unlimited online resources for granted have very specific needs and expectations from their learning environments.... They expect to try things rather than hear about them. They tend to learn visually and socially. They are accustomed to using technology to organize and integrate knowledge" [4].

This new research on learning, plus market pressures on colleges and universities to provide students with access to and experience with the latest technologies, challenges faculty in all disciplines to reflect on their teaching practices, and to consider how to incorporate technology so that it enhances student learning. However, data from the 2002 Campus Computing Survey on course enhancements show that less than 20 percent of all institutions use computer simulations. The same survey reported that only about 20 percent use commercial courseware/instructional resources, about 25 percent use course management tools, and about 35 percent use Web pages for class materials and resources [12].

Where do the Problems Lie?

There are a number of obstacles to winning faculty buy-in for using technology to enhance teaching and learning. They cut across disciplines. First, there are intellectual property concerns such as whether online materials created by faculty may be appropriated by others without due compensation.

There are academic freedom issues. Some faculty perceive that use of technology such as course management tools is being imposed top-down. A related barrier is the tradition of faculty autonomy in the design of their courses. Many, if not most, faculty take considerable professional pride in their teaching and regard it as their unique area of expertise. Some faculty feel that instructional designers, whose job is to help them improve their teaching or their students' learning, have neither the teaching experience nor credentials comparable to their own. Others feel that pedagogical fads come and go.

Also, some faculty perceive that producing and maintaining online course materials is time-consuming and burdensome, and that these tasks will fall to faculty because of inadequate staff support or training. So, there are workload issues. Another obstacle for some faculty is their perception that there are few scholarly studies that show the use of technology significantly enhances teaching and learning. Finally, some faculty fear they risk unfavorable performance reviews should leaving their pedagogical comfort zones result in lower-than-usual student evaluations.

Advice on Strategies, Warnings, and Opportunities

From January through August 2000, the NLII conducted focus groups in order to understand how to win faculty engagement and support for utilizing technology to transform higher education. It identified "twelve campus conditions" that are necessary for winning faculty support, including the commitment and leadership by senior administration [3]. Therefore, an essential strategy for department chairs is to find a champion of technology among senior administrators who can motivate faculty and build excitement about adopting the use of technology without raising concerns about academic freedom.

The chair also needs to capitalize on the technology-savvy members of the department. For example, while young faculty may not have much teaching experience, they often are very comfortable with the use of technology in the classroom and can be drivers of change in the department. The chair needs to find ways of encouraging the junior faculty and of securing the resources they need without threatening the senior faculty. He or she needs to move ahead at a pace the faculty, the department, and the institution can support.

Another strategy is information-sharing. There are several kinds of information that might be useful to share with faculty, e.g., data from student satisfaction surveys on student expectations and demands for using technology, credible studies across academic disciplines which show that the use of technology improves teaching and learning, and labor reports on areas of job growth and on job requirements which highlight proficiency with technology as a necessity.

Another strategy for a chair is to provide a risk-free zone in which faculty may experiment in using technology without fear of being penalized at the time of performance or merit review.

Department chairs must be frank and warn of consequences that may follow from low use of technology in teaching. Alumnae funding for technology might dry up. Students potentially will be shortchanged in their pursuit of the most attractive and lucrative careers. A department that does not meet baseline expectations in the use of technology risks losing student interest.

On the positive side, there are several opportunities available to department chairs for building faculty buy-in. First, there are tangible incentives such as summer stipends or course releases to re-engineer courses, a piece of new technology, and access to top-of-the line classroom space. As well, there are important intangible incentives to offer faculty, such as providing risk-free zones in which they may experiment, acknowledging their accomplishments during a performance review, and publicly acknowledging them at faculty meetings and other appropriate forums.

In the area of technology, the department chair has multiple responsibilities: He or she is responsible for working with the administration to accomplish strategic plans that relate to technology, for serving student needs, and for promoting faculty professional growth and development. Winning faculty support for using technology to promote teaching and learning will require that the chair lead the department in ongoing consideration of best teaching practices and of how best to utilize technology, in developing a departmental technology curricular integration plan, and in advocating for the resources (e.g., technology, release time, training) necessary for carrying out the departmental plan.

Technology in the Classroom

Michael Pearson
Mathematical Association of America

The use of technology by faculty, as a tool for communication, both for administrative and research purposes, is settled. No one realistically expects or wants us to abandon the use of email and electronic typesetting systems (e.g., TeX) that allow faculty to more efficiently carry out their responsibilities. Nor would we want to abandon the use of sophisticated mathematical software as research tools.

On the other hand, the role of technology as a pedagogical tool is far from settled, and it is likely that you will face administrators, faculty and students with vastly different agendas and perspectives on this issue. The problem is not simply convincing faculty that there is some utility to using technology (though there are some faculty who are perhaps too limited in their views), but often convincing administrators that technology is not a panacea, either for pedagogical or financial considerations.

Faculty generally recognize that using technology requires significant investment in resources. There are physical plant requirements: computers and the classroom/lab space to use them. Faculty face an investment of their own time to learn software, then review and adapt existing curricular materials to fit both the available software and their students' needs. Use of technology requires classroom time to teach students how to use the software. How many of us have much experience teaching students how to use technology? The department head must be able to effectively represent these realities, both to administrators and faculty who may be unrealistic in their push to adopt the use of technology across the curriculum.

Administrators may have unrealistic expectations regarding costs of incorporating technology into the curriculum. Their reasoning is often along the following lines: "Let's build a big computer lab, put all our developmental courses on-line, and lower our overall cost of instruction." I am not aware of any institution that has been able to generate significant cost-savings using this strategy. At best, when used appropriately such tools may prove to be cost-neutral, but I'm not even convinced that's possible. Technology may be a great tool when used appropriately, but will likely cost more money.

Another issue we must face is the use of competing technologies across campus. For example, business schools are not likely to be interested in adopting computer-algebra systems in their courses, usually preferring spreadsheet applications. Engineering schools may prefer *MATLAB* or some other numerical plus graphical package. These choices are made for good reasons. Yet mathematics faculty usually prefer *Maple* or *Mathematica*. Is it realistic to expect students to put in the time to learn different packages for different courses? I would argue that students will be very unhappy, and not particularly cooperative, if there is a fragmented approach. Department heads must take some initiative to coordinate efforts to adopt technological tools with other campus units.

Once a plan for incorporating technology into particular courses has been put in place, support for those technologies for both faculty and students is essential. For faculty, some mechanism for sharing resources must be set up. It is not appropriate that every faculty member reinvent, or re-adapt, technology-based curricular materials for their courses. Moreover, this sharing needs to take place in a way that preserves some consistency across multi-section and prerequisite courses. Existing problems of course coordination can be exacerbated by course realignment, often caused by new technologies.

Appropriate support for students must be provided. A group of students in a computer lab, frustrated with a software-syntax problem and with a project due the next day, is an unhappy group of students. If help is not readily available (e.g., from a knowledgeable lab assistant), the department head will likely soon hear their complaints. Worse, if the campus or lab network crashes and leaves the lab facilities unavailable, students may appear en masse to vent their frustration, perhaps understandably so. I think it is because of this time-sensitive nature of the support required for successful implementation of a technology-based program that it is difficult to realize significant cost savings. It is simply not realistic, in the context of a typical course, to log technical problems with the campus computer support office and then wait some days to get a response. Moreover, it is not realistic to expect campus support personnel to provide adequate technical support for mathematical software, thus the need for significant investment in support staff in the department. This may be done using graduate students, or even advanced undergraduates, but will require significant faculty oversight; again, an investment of departmental resources.

Finally, an issue that spans multiple areas: assessment. How do we evaluate the effectiveness of the use of technology in the classroom? Are we using technology to reinforce traditional types of knowledge, or are we instead preparing students to do different things altogether? In my experience, these questions are usually not addressed in a meaningful way and are often a source of difficulty inside the department, but may prove to be an even bigger concern as we deal with administrations increasingly focused on issues of accountability.

Reviewing my remarks, one might assume that I am opposed to significant use of technology in the classroom, if not a downright Luddite! The opposite is true; as a faculty member, I have always supported incorporation of computer-based work, at least in courses at and above the level of calculus. I have spent many hours installing and troubleshooting software, and even done carpentry on weekends to retrofit department space to accommodate new computer equipment. At some time or another (in fact, lots of times) we encountered all of the problems I describe. Nevertheless, I am still in favor of bringing new technological tools to our students, and feel that it's essential to do so, facing and overcoming the problems with as much wisdom as we can muster.

For example, faculty who are motivated to incorporate technology-based components into their classes can offer seminars for their colleagues, introducing the software and demonstrating what they do in their classes. Faculty can also be encouraged to post materials publicly, either on a website or an accessible network drive, both to help stimulate new ideas and promote general awareness of what is going on in various sections of what are, after all, courses "owned" by the faculty as a whole.

When discussing the use of technology with the dean or other administrators, the need for adequate support in order to effectively incorporate technology into the curriculum should be kept front and center. Stress the ways in which your department feels that these new tools will improve your ability to meet instructional objectives, and why you think it is worthwhile to make the required investments.

Of course, when you ask for something new, you probably can expect to be asked what you might be willing to give up. Be prepared for the question; know what your department can afford to relinquish or reassign. Don't be caught off-guard and leave your dean in a position to choose a reallocation strategy for you.

Program Review

Donna Beers
Simmons College

On a periodic, usually five-to-seven year basis, many institutions require departments and programs to undertake internal and external reviews of their offerings for purposes of curricular renewal. Not only is self-examination and renewal presented by deans and provosts a good in itself, but surveys of prospective undergraduates and their parents show that the top selection-drivers for prospective students and their parents include the quality of the desired major, employment opportunities after graduation, and preparation for graduate or professional school.

Academic institutions exist in a competitive and fast-changing environment. They are challenged to market their programs in ways that respond to market forces while remaining true to their core missions. The main 'product' that an institution has to offer is its curriculum.

Mathematics departments play a special role relative to the college curriculum: They serve students majoring in mathematics and in new interdisciplinary majors such as financial mathematics and bioinformatics, they help prepare future teachers, they serve students from a host of departments and programs that require statistics and calculus, and they are the stewards for an institution's quantitative reasoning requirement. For these reasons, a program review for mathematics should include environmental scans to ensure that the curriculum is current and competitive and to ensure that it fits with and supports the institutional mission. Very important, departments must measure the quality of teaching and learning to ensure that they are successfully delivering the curriculum.

Departmental Buy-in for Program Review: Where do the Problems Lie?

Formal program reviews are a fairly recent phenomenon at many institutions. For department chairs, motivating faculty to participate in carrying out program reviews can be a major challenge. Objections include the perception that everything is okay, the perception that program review is being imposed from top-down, the absence of positives such as incentives or rewards for doing this work, and the presence of negatives such as the beliefs that program reviews entail a lot of work, they are time-consuming, and they're an addition to all the regular work. Perhaps the greatest objection is the question: what good will it do?

Advice on Strategies, Warnings, and Opportunities

To ensure that department chairs are successful in leading program reviews, the administration must make program review an institutional priority, provide a rationale, and give a timetable for each department to be reviewed. The dean of each college must work with the college or university-wide curriculum committee to develop guidelines for departments and programs to follow in carrying out program review.

Also, the administration can provide department chairs with carrots to use as sweeteners for motivating faculty to contribute to curricular renewal. These include: faculty release time or summer stipends to re-engineer a course; faculty release time or summer stipends to prepare and teach new courses in areas of potential growth that have been identified in the program review process; the opportunity to pilot a new pedagogy in a risk-free zone (to protect faculty from the possibility of receiving lower-than-usual student evaluations because they have ventured from their comfort zone); access to premium classroom and lab space for faculty who commit to utilizing new technology in their teaching; and formal recognition during the merit review process for contributions to the program review.

In addition to carrots for individual faculty members, central administration can offer incentives to the department as a whole. These include informal, public recognition (e.g., at college-wide faculty meetings and other appropriate forums) for carrying out a successful program review; new equipment and resources for departmental initiatives which will enhance teaching and learning; and, new hiring slots to acknowledge the department's success in growing its majors. To encourage cross-disciplinary initiatives, the administration must be willing to make joint appointments and pilot and fund new initiatives.

On the other hand, department chairs need to make faculty aware of administrative 'sticks' that may be applied should a department not carry out program review. For example, courses that have not been offered for several consecutive years may be automatically dropped, and/or courses that have had consistently low enrollments may be periodically reviewed by the college-wide curriculum committee for possible recommendation of being dropped. Also, departmental requests for new resources, material or personnel, may go unheeded and/or faculty sabbaticals may not be automatically approved. In addition, faculty might receive low (or no) merit increases.

An Example of a Success

Our departmental program review was overall a very good experience. Positive outcomes included the following:

- We analyzed course enrollment trends and uncovered declining enrollments in a couple of "bread and butter" courses. We became ambassadors and cultivated new audiences for mathematics. We developed new courses that have helped build enrollments and reverse declines.
- We benchmarked our departmental majors and uncovered some gaps. The environmental scan, together with input received from alumnae focus groups, led us to update and strengthen our majors, enhancing their depth and breadth.
- Our reviewers, both external and internal, affirmed the high quality of mathematics instruction and the exceptional devotion of faculty to their students. This helped to boost departmental morale.
- Responding to the recommendations of reviewers, the administration provided new resources, including new technology and a new tenure-track hire.
- Very important, we renewed communication and collaboration with other departments to develop new courses to serve their majors.
- The department received public acknowledgement and kudos for carrying out an exemplary program review.

Faculty Issues

Faculty Review
  Connie Campbell

Merit Pay
  Bruce Ebanks

Hiring and Firing Staff and Faculty
  Bruce Ebanks, Jimmy Solomon, and
  Tina Straley

Mentoring Faculty
  John Conway
  W. James Lewis

Dependence on and Culturalization of Part-Time and Temporary Faculty
  Norma Agras
  Donna Beers
  Catherine Murphy
  Connie Campbell

Resolving Conflict within the Department
  Norma Agras
  Connie Campbell

Managing Conflict with the Dean, Provost and Other Departments
  Connie Campbell
  Jimmy Solomon

Faculty Review

Connie Campbell
Millsaps College

At Millsaps College, all full-time faculty are required to write an annual review of their performance detailing their contributions and growth within the areas of teaching, scholarship, and service to the college. A copy of this self-assessment goes to their department chair and also to the academic dean. Department chairs are then asked to write an annual review of the full-time faculty within their departments. This review is somewhat a response to the faculty's self-assessment, but it is also meant to help the faculty member gain perspective on how their department chair perceives their performance in each area. The department chair's evaluation is distributed to both the individual faculty member and to the academic dean. The academic dean then works together with an elected review committee to rank the faculty in each of the three areas and to determine an overall ranking of the faculty member. Faculty who fall into approximately the upper third of an area are recognized with a merit pay increase for their contribution. Additionally, the overall ranking of a faculty member in comparison to their peers determines their annual raise.

On the positive side, this system provides faculty with a regular and systematic way to assess their development in each of the areas of teaching, scholarship, and service. In this process, faculty are encouraged to reflect on both their successes and their failures, providing a continuing dialogue on their commitment to growth. Additionally, this process allows for input from the department chair, who is perhaps best positioned to assess the contributions of mathematics faculty, while also providing a normalizing effect in that all faculty are ultimately evaluated by the same person.

Unfortunately, within this system, faculty often see their department chair as a person who is writing them a letter of recommendation for a raise, rather than a person who is actually evaluating their performance. They perceive that the true performance evaluation comes from the dean/committee level. Due to this misperception, many faculty and perhaps even chairs do not use the department level and self-evaluation processes effectively. Instead of seeing the process as a means of constructive feedback intended to support and guide, faculty often assume a defensive stance, thereby limiting their self-reflection and critique of their own performance as well as benefiting less from that of the chair. Another obvious drawback is that faculty who are more willing to write a compelling case for their commitments typically fare better than those who may do more, but are not comfortable with, and/or willing to invest the time into, explaining the greater implications of their contributions.

Particularly in a system where the variation among raises is small, merit pay is largely symbolic in that it reflects the belief/value that performance should count. However, merit pay, particularly when it is based on a comparative assessment and not clearly defined benchmarks, leads to competitiveness and self-serving goals. Departments (or colleges as a whole) should find other ways to reward their faculty. We should not rely on merit pay and peer reviews to motivate faculty. Rather we should find every opportunity to recognize their efforts and give them feedback and encouragement. Faculty will be motivated to perform well if they have a high level of job satisfaction. This is where the role of the department chair is particularly critical. Chairs should provide their faculty with attainable goals, substantive feedback, recognition for accomplishments, and the opportunity to be involved in decision-making.

On the other hand, there are certainly problems inherent with systems that do not utilize merit pay. When performance is not connected to raises, the chair's incentive to provide a serious review as well as the faculty's incentive for self-reflection and growth is dramatically reduced. While I would not argue for a merit pay system, it certainly has the advantage that it promotes a vested interest for a more thorough review.

Merit pay or no merit pay, it is imperative that faculty be given feedback on their performance on a consistent basis. For the evaluation process, department chairs should do everything they can to gain a valid and full picture of the contributions of each member of their department and to help the faculty develop to his or her full potential. Chairs should: read student evaluations; plan peer visits to class; engage faculty in discussions over pedagogy, course materials, and course content; facilitate discussions concerning research accomplishments and goals; and engage in discussions about what types of service to the college would be commensurate with the individual faculty member's interests. These are all ways that the chair can build a better understanding of the contributions individual faculty members are making, while also encouraging commitment to the overall goals and mission of the college. Faculty who believe that they are valued will perform better and will strive to improve.

Additionally, chairs should meet with each departmental faculty member and discuss his or her review, allowing the faculty member not only the opportunity to talk about any questions or concerns he or she might have, but also to facilitate discussion of the faculty member's contributions and areas for growth. When done correctly, reviews should offer no big surprises. Faculty should be adequately informed about job expectations and should be given on-going feedback of their performance. Faculty review at its best is a continual dialogue that has positive implications for collegiality, job satisfaction, and commitment to the institution.

Merit Pay

Bruce Ebanks
Mississippi State University

This issue stands alongside hiring and tenure/promotion decisions as being one of the toughest for a department head. The problem, of course, is that each and every faculty member considers himself or herself to be excellent and therefore worthy of the largest possible amount of merit pay. The department head sometimes must do his or her best imitation of Solomon to make difficult judgments about relative levels of performance and merit. How does one accomplish this task?

Success requires a strategy employing several ingredients. First, make sure the criteria used for faculty review properly reflect institutional and departmental goals and priorities. This is necessary to assure support from both the higher administration and the department members. The criteria may have to be modified from time to time as priorities change or if a determination is made that the criteria do not adequately reflect priorities. Agreement on these criteria requires a lot of communication in both directions. There should be discussion within the department about how departmental priorities tie in with institutional mission and goals, as well as how well the faculty review criteria connect with those priorities. There also must be some discussion between the department head and the dean about the departmental priorities and how they are connected with advancement of the institution.

Second, both the faculty review criteria and the review process should be communicated as clearly and understandably as possible. There should be a department document that addresses this issue, and this document should be re-examined periodically to see whether it captures the essence of what the department and institution truly value. If the document does not reflect the mission and strategic plan of the institution and department, then it is time for some changes. The review process should be communicated clearly to faculty members. This applies especially to new members of the department. Everyone should know what the ground rules are for faculty review and merit decisions.

One of the greatest difficulties in this process lies in attaching weights to the various criteria. Each person tends to feel that what he or she is doing is most significant. Several things may help the department head here:

- As far as possible, it may be desirable to describe quantitative measures associated with each criterion used for review.
- The department should have some basic agreement as to the ranking of criteria. Examples may include such things as: (a) published articles in refereed journals are more important than conference proceedings papers; (b) chairing an important committee carries more weight than refereeing papers, etc.
- During the faculty review process, it may be useful to incorporate specific goals for each faculty member. The faculty member should be encouraged to propose such goals for the coming year, subject to agreement by the department head.

Some departments and some administrators tend to overemphasize item (1) above. It is no good to measure quantity without also assessing quality associated with each contribution. I know of one department where a scheme was devised by which the numbers of publications, presentations, grants, etc., were entered into a formula by which a research rating number was calculated for each faculty member. The idea was to have a purely objective rating scheme. The result was of course that the rating scheme encouraged people to try to do the least amount of work necessary to achieve the highest possible rating. The moral of the story is to try to make your review system as objective as possible, but not more so. There must always be some room for value judgments.

Most departments have multiple missions, and individual faculty members have different strengths. It is desirable to reward people for whatever they are doing, as long as it is helping the department, program, and/or institution. So the review and merit system should build in rewards for contributions in all areas of the department's mission. The department head should view the department as a team, and try to ensure that the team meets its goals. It is not necessary for each member of the team to do the same things. The department head should try to identify and make use of the particular set of strengths of each member of the department, always with a view toward the overall department goals.

On the other hand, if a determination is made that some contributions should be valued more highly (for instance, because of institutional or unit mission), then the department should recognize that by providing proportionately greater rewards for those contributions. For example, if the department is located in a major research university, then greater emphasis should be placed on research contributions. Of course this does not mean that contributions in teaching or other areas should be ignored.

Finally, merit pay must be tied closely to the faculty reviews. Generally, someone with a higher review rating should get a larger merit increase. If this is not the case, then the reasons why should be known and accepted by the department. Such a situation could arise if the faculty review instrument calls for separate reviews in each of several areas—say, teaching, research, and service. But perhaps the institution or department values contributions in these areas with different weights when it comes to merit pay decisions. This is all right as long as the process is rational and understood by the department.

Hiring and Firing Staff and Faculty

Bruce Ebanks
Mississippi State University
Jimmy L. Solomon
Georgia Southern University
Tina H. Straley
Mathematical Association of America

Hiring and firing are intricately connected because the best way to avoid firing people is to make good hires to begin with. Even under the best process, however, you don't have a crystal ball, and there will be times when you have to terminate faculty or staff. Below, we discuss the all-important hiring process, what you need to do when problems arise that may result in termination, and suggest a process for termination. This section is not about termination that follows from a negative tenure decision. That decision is not solely the chair's to make, but one to which the chair has input as part of a well-developed, statutory process at your institution. The discussion on hiring and firing of staff is integrated with that on faculty or separated when there are significant differences. There are many similarities but also major differences.

Hiring Faculty

Hiring faculty is probably the most important issue for departments and one of the most difficult for department heads. Getting the right people in the right positions is critical for the success of the department. Many academics have a tendency to be too passive when it comes to recruiting and too lenient during the probationary period of a tenure-track faculty member. In addition to the obvious fact that good hiring is fundamental to improving a department, good hires help to minimize the headaches and stress of a department chair. You want to hire faculty who will contribute to the improvement of the quality of the department. Never fill a tenure-track position just for the sake of filling the position.

Defining the position: The hiring process starts with permission to advertise for a position. Already a big decision must be made. Do you specify a desired area or areas of research, or do you seek the "best" person you can get regardless of area? Clearly answers will vary, but one should consider the question carefully since the answer charts a course for the search process that will be extremely hard (if not impossible) to modify later. Often there is a particular programmatic need for expertise in a certain area or areas, and this will determine how the ad is written. If this is not the case, it still may be useful to set certain priorities up front so that one doesn't have to compare apples and oranges later on. It is exceedingly difficult to get a search committee, much less a whole department, to agree that professor X in area Y is better than professor A in area B. On the other hand, it is sometimes useful to write the ad in such a way that area Y is preferred but applicants in other areas are also welcome to apply. If the search comes up empty in area Y, then the department may want to try to hire in area Z that year, then return to area Y the following year. (That is assuming the department will have another position open the following year!) This part of the process might be done before a position is secured or it may be done later with the input of the department or the search committee.

Once the programmatic area is decided, the next step in the hiring of faculty is to have a clear idea of the specifics of the position. This implies that there is a clear understanding of the duties, salary range, and resources needed to support the new hire; e.g., if the salary range which is available does not allow for a competitive opportunity at the full professor level, it may allow for a very competitive salary at the senior associate professor level. If this is the case, advertise for an associate, not for a full professor. If the position will result in a department member whose job description represents a significant change from the existing "culture" of the department, then that should be understood and supported by the faculty from the beginning of the search.

Solicitation and recruiting: The next stage in the faculty hire is one that many departments ignore—active recruiting. Many departments simply put out their ads and wait for the applications to roll in. But there can be a considerable advantage in actively seeking applicants. Perhaps potential applicants do not consider applying for your position because they don't know enough about your department or they have a mistaken impression of it. Sometimes personal contacts at other institutions can help you find applicants who would not otherwise consider your position.

The search process: Appoint a "blue ribbon" search committee. A search committee is the department's ambassador to prospects. Appoint individuals who have a broad network in the mathematical community or at least in the area in which you are trying to hire. Such breadth on the part of a committee member allows for a better opportunity for more meaningful input in trying to hire the best individual possible for the position.

Next comes the difficult **winnowing process.** Many departments can only afford to bring in two or three candidates for on-campus interviews for a position. How do you get from hundreds down to just two or three? Of course the first cut has to be done on paper when you have a large number of applicants. Ideally, the search committee will create a short list of perhaps ten or twelve for telephone interviews. It may also help at this stage to talk to references, especially if there is some doubt or ambiguity in the reference letters. When speaking to candidates, it is good to have more than one person from the department involved. (You can use a speakerphone.) The telephone interview is especially critical for considering foreign-born applicants if facility with English may be an issue. It is very helpful to have several experienced department members converse with applicants before making the final selection of the few to be interviewed in person. There is no point in wasting time, yours and the candidate's, and money by bringing in someone who turns out to be unacceptable or just a bad fit.

The on-site interview should be very carefully orchestrated. You are reviewing the candidate to determine if this is someone you want to add to your department. The candidate is reviewing the department and the institution to determine if this is a place in which he or she fits and wants to spend a significant part of his or her career. The interview should include meetings with whatever groups are important including students, faculty from other departments (depending on the position), the dean and other administrators. There should be time for informal discussions as well as more formal meetings. There should be a presentation that covers whatever is important in the position, primarily scholarship and teaching.

At the campus interview stage it is essential to get feedback from as many members of the department as possible. You may have a candidate rating form that is given to all faculty members. Encourage each faculty member to visit with the candidate and to write comments on the rating form. A colleague may see a side of the candidate that others don't. For the same reason it is very helpful for the chair or the search committee to chat with individual faculty members about their impressions of the candidates. Then you will usually have a pretty good idea of the ranking of candidates before the department meeting at which this is discussed.

Hiring Staff

The process for hiring staff is most likely determined by your university. Depending on the staff position you may be advertising locally in the newspaper, on campus, or in trade journals. Many of the rules above still apply. You should involve faculty or staff in the selection process by having them serve on a search committee. You should winnow the selection down to about six to ten candidates based upon the applications. Then conduct telephone interviews. There may be tests of job skills that the institution gives and you may wish to augment these tests based upon your specific needs. Invite two to three candidates for on-site interviews. Interviewees should meet with faculty and staff and any other groups with whom they will interact. Just as with faculty, do not fill the position out of desperation. Be patient and wait for the right person for the job.

The whole process is certainly imperfect. Even after the most careful search process, the new faculty or staff member may not work out as anticipated.

Dealing with Problem Performance

Never ignore problems hoping they will go away. They only get worse. You must act quickly. If you realize in the first year that a faculty member is not what you thought and there are major problems that cannot be remediated, take action and do not rehire for the second year. This same warning holds for the first three years. After that, it is harder to terminate without going through elaborate procedures. Staff positions usually have a short three to six month probationary period. If you have any reason to believe in that time the person is not right for the job, take action and terminate the person during this period. It is far easier to do now and much harder to do once the person has become a fixture in the department. However, you don't want to be too hasty if there is a chance the person can work out. After all, you've made an investment in hiring. There are steps to take that will help you determine what course of action to follow.

Mentoring: If a new hire is having problems, you should try to help. The person may have much to offer if some adjustments could be made and you will find that you have a valuable addition to the department. No one likes instability, so salvaging a position is better than termination and a new search starting over again with someone you do not know and about whom there are no guarantees. You should have a procedure in place to provide peer mentoring and to acculturate new faculty and staff. As chair, you should be helping them adjust to, and be successful in, their new jobs. One approach that might work with one person may be a total failure with someone else. To some extent, the help that a person needs depends on that person and what the difficulties are.

Keeping a paper trail: You never know where problems may lead. You always hope that they will be solved and forgotten. However, you need to prepare for any outcome. You must create and keep documentation of everything that is presented to you and that you do to solve the problem. You must especially have records of your communications with the faculty or staff member who is experiencing problems. If you get complaints that are written or oral, share them, perhaps anonymously, with the person, keep the original correspondence or your notes of oral communications in the personnel file and inform the person you are doing so. The individual should have the opportunity to write a response to anything put in his or her file. You should not keep anonymous accusations for which you have no verification. If you counsel the faculty or staff member, write a memo afterward summarizing the discussion. Have the person sign the memo to indicate receipt and keep a copy in the file. Keep records of any interventions the person goes through. Make sure to address problems in the regular reviews you write. If the problem seems small, say that, but address it none-the-less. If it grows, you have documentation that the person was informed.

Giving fair warning: If problems become serious and you see that they could lead to termination, give warning as early and as clearly as possible. If the person is beyond a probationary period in which you can terminate without explanation, let the person know that the behavior causing the problems may lead to termination if not corrected. Give the person a chance, even a timeframe, in which to make adjustments.

If one does the hiring process well, then the probability of later having to terminate that faculty or staff member is reduced but not eliminated. Terminating a non-tenured faculty member is less likely to result in litigation than terminating a tenured member of the faculty. Termination of a staff member may also lead to litigation. The best advice is to maintain good documentation, and make sure that established procedures are followed. As with all personnel matters, your discussions during the process should be limited to those with a "need-to-know." You may need to consult with someone in your Human Resources or Personnel Department about the review process for staff. They can give you a better idea about what is expected in the way of warnings before an unsatisfactory staff member can be terminated.

Firing

Untenured Faculty: The decision to fire a faculty member is a very difficult one. Usually, the department has gone through a long search process, with many person-hours and many dollars expended along the way. The department may be reluctant to consider letting that person go just a few years later. Remember that it gets harder and the investment gets bigger with each additional year.

Often departments will give a tenure-track candidate the full six years of the probationary period. They hope the person will grow into the kind of colleague they want. But what if, in spite of all your best efforts, the person gives little indication of becoming that hoped-for colleague? If you wait until the end of the probationary period, the whole department has to agonize over the tenure decision. If a serious effort is made to mentor and develop a young faculty member for a few years, and if it is clear that the person is not making satisfactory progress, then it is preferable to terminate the faculty member before the time comes for a tenure decision. Certainly it is in the best interest of the department, and most likely it is also in the best interest of the individual affected. Perhaps the department, institution, or location is just not a good fit for that individual. That person may be happier and more productive in a different job and/or environment. Is it not better for that person to look for a more suitable situation sooner rather than later? Finding another position, before the up or out decision, removes the stigma of a negative tenure decision from the faculty member's résumé.

Do not be too disappointed if individuals change their minds during the process leading to termination. Nobody likes firing and it often happens that some people will change their minds when the going gets rough, for example if they are asked to testify at a hearing.

Staff: In the case of staff, your main contact will probably be with Human Resources. However, it is still important that you keep the dean informed as you proceed. The procedures for terminating staff are generally different from those of faculty. In fact the process may be taken over by your Personnel or Human Resources Department.

The issue of firing staff is perhaps less difficult because there is not an issue of tenure. Nonetheless many of the same considerations are present, and there is an expectation of continued employment for long time employees. As stated above, it is difficult to make the decision to fire a person with whom you have worked. Yet sometimes you may need to make that difficult decision for the health of the department. If you have followed all procedures established for your institution, you have given the staff member all reasonable opportunities to make changes, you have consulted with HR or Personnel, and you have the go-ahead to fire, then you can do so with confidence in your decision.

Tenured faculty: This is almost impossible unless the faculty member is guilty of illegal acts, gross incompetence, or extreme insubordination. Many systems and institutions have initiated post-tenure reviews to deal with this problem. However, even these policies and procedures rarely lead to termination without going to court. If you are having problems with a tenured faculty member, you need to find some other way to get the person to change or you need to change your expectations and live with the situation. Sometimes the person can be transferred to a different department or a different position.

Should you be faced with terminating a tenured faculty member, you will come to realize that there is no such thing as too much documentation. You will be glad that you followed the advice above and started keeping the documentation early. In addition, you should discuss the problems with your supervisors and they might be able to advise you or even reassign the person. In particular, you should keep the dean fully informed of your intentions and the procedures that you are following. You will have the support of the appropriate faculty and administrators —those with a "need-to-know"—if you have kept the records and have kept them informed.

Firing is always traumatic for everyone involved. As an administrator, it will be your hardest and most disturbing task. However, you and your department would not want you to be the kind of administrator for whom it is easy.

Mentoring Faculty

John B. Conway
University of Tennessee

The most crucial part of mentoring is fostering the beginning of a career. To get a new faculty member off to a good start is the responsibility of the entire department, but the leadership for the task usually falls to the head as the principal in charge. Just as it takes a village to raise a child, it takes a department to facilitate the transition of a new faculty member to a respected and productive colleague. A department head must also be conscious of mentoring more senior faculty as they progress through their careers, encounter problems, and pass through a span of perhaps thirty-five years in your department.

Of course the most important aspect of mentoring junior faculty is making sure they do well in the classroom; teaching is the aspect of the profession that graduate schools neglect the most. Now the first thing to be aware of when you or a colleague mentors a new assistant professor in teaching is that there is more than one way to be a successful teacher. There are some basics of which everyone should be aware: don't stand in front of what you have written; never make fun of your students; etc. But how you present a lecture, how much detail you provide, what to expect on a test are things that can vary widely and still result in success.

Junior faculty may be anxious to try a variety of innovative teaching strategies but may have little understanding of the complexities inherent in them. Principally among these is the use of technology in teaching and learning. You may find an inexperienced teacher taking on a new and untested teaching strategy, delivery system, or course materials for which there are no experts in your department. Don't try to impose a style on the assistant professor. A good use of student evaluations is to see how the various practices are affecting students. (By the way, the practice by almost all colleges of using student evaluations as the sole means of evaluating teaching is one of the travesties in higher education; another example of number abuse and abuse of the profession.) It is a good practice to go over these evaluations with colleagues, junior or senior, and try to discover what caused both good and bad responses. Teaching evaluations may be particularly helpful to the faculty member trying out new approaches, teaching methods, or curriculum for which there are no tried and true maxims on which to rely.

Class visits are recommended, but again I would try to avoid having a junior colleague regard these as judging his/her performance. This is one reason why it might be best to have someone besides the head as the individual's mentor. It's also a good idea to have the assistant professors discuss their exams, major assignments, and other materials they create with their mentors. A practice I liked to follow when I was head was to suggest five faculty members whose classes the junior colleague should visit. Pick the better teachers and let the new assistants see examples of what you consider good teaching. Encourage them to discuss their visits with their hosts. Besides fostering good teaching it enables some of your senior faculty to become better acquainted with the new ones. Often the senior faculty members learn from the observers and gain insights to help them improve their own teaching. In all cases, the ensuing discussions are interesting and benefit all parties.

Besides teaching you should mentor the research of your new faculty. Clearly you aren't going to suggest research topics or do what a thesis adviser did. But new PhDs have gaps in their understanding of the research world. When I started at Indiana University I still remember assuming I had to wait to be invited to a conference. My thesis adviser, who did a very good job directing my dissertation, never told me that no one would think it inappropriate to contact a conference organizer and ask to give a talk. There are also questions about where to publish, how to give a research talk, how to write a grant proposal, whether to write two short papers or combine the two for a longer one. Yes, you would hope they would talk to their thesis advisers about these things, but they might not for any one of many reasons. Indeed they may want your blessing on continuing to consult their advisers. Some questions might better be answered by an expert in the research area, but make sure they talk to someone. Even if you aren't an expert, your point of view might still be appreciated.

A new faculty member has not only joined your department, he or she has also joined the institution, the local community, and the community of mathematical scientists. A new faculty member is not only learning how to teach and how to continue research independently, but also trying to figure out the expectations for fitting into these communities. For many junior faculty members there are family, social, and local community adjustments to be made as well. Again, the faculty's best sources of information are the department head and the individual's mentor. Junior faculty members may volunteer for too many department tasks in order to please or too few because they don't recognize the importance of this work. They don't have any idea what is reasonable unless they are given direction of what is expected. In fact I always told my female colleagues that it was part of my job to keep them off too many committees. Universities have an insatiable appetite for women and minorities for the campus committees. Too much of this will skew and perhaps undermine their careers.

Outside the institution, the demands and opportunities are just as bewildering. What conferences should he attend, where should he present his work, what committees should he volunteer for, which professional societies meet his needs and interests? You should help the new faculty member choose institutional and mathematical community activities in which to participate. But be sure to encourage them to join. I have advised all my new colleagues to become a member of at least two mathematical organizations. These organizations represent us and have to be supported.

A good way for a new faculty member to become a part of the professional community is to participate in MAA's Project NExT for recent Ph.Ds, your MAA Section NExT, or Project ACCCESS for new two-year college faculty members. Project NExT and the Section NExT programs are supported by a number of professional organizations in the mathematical sciences as well as private foundations who value the success the program has in assisting new faculty at this critical career transition point. Project ACCCESS is a collaboration between MAA and AMATYC supported by the ExxonMobil Foundation.

The career of a professor is long and it's a service to the individual, the college, and the profession to help new faculty to start correctly. Remember that when someone fails to get tenure, they are not the only ones who have failed; so have you and your department.

A new faculty member may have a mentor for only the first year or two. However, if the relationship between the mentor and the junior colleague is a good one, the mentoring may continue throughout one or both careers. In most cases, however, the formal mentoring will end by the time the junior faculty member has completed one to three years. After that time, the department head will assume the role of mentor. The head can be the mentor to all faculty in the department and is the only person in the position to do so. The level of mentoring will vary with the experience of the person and that person's specific needs. You may meet with new members of the department frequently and then continue to confer with faculty members with decreasing regularity that is a function of their number of years in the profession and in the department. The minimum of a once-yearly review of all faculty is a necessity and is often required by the higher administration. If taken seriously and done well, the reviews are more an opportunity for assisting the faculty members along their career trajectories than they are meetings to determine merit raises.

As a career progresses new problems arise for the faculty member and the head has a continuing role to play. Yes, it can be awkward giving advice to a senior colleague, especially if they are older than you. But, if done right, it will be appreciated and will help your department run more smoothly and maintain a more collegial atmosphere.

A fact of academic life is that age frequently quells the fire for research and scholarship. Even when senior faculty continue doing research longer than the norm, the quality of that research is likely to diminish. To say that research is a young person's game is to overstate the case, but this contains a grain of truth. There may be a clearly discernable watershed event to mark this, like losing a grant. The more likely scenario is that there will be gradual erosion over a period of years. Many have adjusted well to this change in status, but many have not.

The head has to play a delicate role in this situation. You don't want to prematurely declare that a change has taken place. Like all such matters, there are bad as well as good ways to approach it and which is correct depends on the individual. Moreover the symptoms are often not clear. Having annual conferences with your faculty where you discuss their work will, over the years, frequently reveal such problems. Even if you are a short-term head, the fact that faculty prepare for such conferences may help them evaluate their progress and recognize the change. Like all good psychologists your main job is to listen and provide a frame of reference. Frequently you have to "give permission" to them to change what they do.

If someone has worked for 20 years in a certain way, she may be unsure how changing the direction of her professional work will be received by you and her colleagues. Will people think less of me? Will some new activity be valued? Will my raises diminish? Will decreasing my research and increasing my involvement in educational matters reduce my influence in the department?

After recruiting and tenure decisions, the most important duty of a head is to ensure that every colleague has a satisfying professional life. Do this well and you will be judged a success. Do it poorly and you are likely to be thrown out of office. Mentoring at all levels is a vital component to being successful in this job.

Mentoring: The Chair's Role and Responsibility

W. James Lewis
University of Nebraska-Lincoln[1]

One of the pleasures of serving as department chair for an extended period is the opportunity to watch as the faculty you have hired grow and develop as professionals, reaching the point of earning tenure and promotion to the rank of associate professor or professor. When asked, we quickly identify hiring and leading the tenure and review process as two of a chair's most important responsibilities. Equally important, but subject to overlooking, is the mentoring of faculty, especially during the probationary period.

We will define mentor as a wise and trusted advisor and mentoring as the process of serving as an advisor or role model who shares professional or personal experiences and knowledge, enabling a less experienced colleague to grow and develop into a successful employee. Central to the relationship is the openness and trust that develops if the junior faculty member believes that the mentor clearly supports them and wants them to be successful.

A good department chair (and in fact most senior faculty) will recognize that it is important to provide successful mentoring for new faculty. After all, hiring is both time consuming and expensive and you want your new faculty to be as successful as possible. It is the nature of our profession that most of the new faculty we hire come from another institution, another state, and often another country. Even if that new hire is incredibly bright, the process of adjusting to the culture and expectations of your department, your institution, your city, etc. is difficult and involves much uncertainty. It is in your own self interest to help new faculty make the transition from new hire to experienced senior faculty.

From the opposite point of view, the new faculty member should welcome any offer of assistance or mentoring that comes their way. After all, there is much on their plate—they want to achieve quite a bit during the time they are untenured faculty members. In particular, they want to become successful mathematicians; they want to earn tenure and promotion; and they want to enjoy "the good life" that comes with being a successful professional.

In offering advice about mentoring and mentoring programs, it is important to offer a word of caution. One size does not fit all. The needs of two untenured faculty members may be very different. At different points in a faculty member's career, their needs may be very different. And the types of mentoring that a faculty member needs at any one time may vary and may call for different mentors. It is the department chair's responsibility to size up the ways in which each faculty member may need mentoring, to make certain that the needs of each new faculty member are met, and thereby to enable each faculty member to be as successful as possible.

In considering this last piece of advice, it is beneficial to consider the different aspects of one's life that might benefit from the information or advice of a trusted friend or colleague. Your new faculty member may be interested in social issues such as "How do I meet people with similar interests?" or "Do you know anything about the quality of the local public schools?" They may need help with personal issues from those as serious as divorce or caring for an elderly parent to tasks as pleasant as being an interested listener when a colleague wants to talk about a movie they saw, a nice local restaurant, or the school's basketball team.

Cultural issues come in more than one type. A faculty member new to your city may want to know how to make contact with the local Chinese community or where to purchase certain kinds of food. As a member of the department, new faculty need to learn about and become part of the culture of the department. This includes issues such as whether to work with your door open or closed, whether someone will notice (and criticize you) if you don't attend a departmental colloquium, and what is expected in terms of how faculty interact with students.

Becoming part of a mathematics department includes understanding life in the department and the institution from many professional and regulatory points of view. Most new faculty reach a point where they ask the age old question, "How do I balance my research with my teaching responsibilities?" They need to have answers to very serious questions such as "How much research is expected of me by the time I am reviewed for tenure?" and they need to have advice regarding which service responsibilities and how many a junior faculty member should accept. New faculty often need help becoming outstanding teachers and they need to know whether, and

---

[1] In 1998, the Department won an NSF-sponsored Presidential Award for Excellence in Science, Mathematics and Engineering Mentoring in recognition of the Department's success in mentoring women graduate students.

how much, professional travel is expected. Most important of all, untenured faculty need a very clear understanding of how they will be reviewed annually and when they are considered for promotion and tenure.

Once we appreciate the many needs that a faculty member may have, we also realize that there are many types of mentoring and that it is seldom that one person can be "all things" to even one person, let alone "to all people." Because there is no one else who clearly has this responsibility, the chair should accept that it is his or her responsibility to put in place the different types of mentoring that a faculty member may need.

Ideally, departments that have developed a strong, supportive culture will have faculty for whom it is natural to be both a friend and peer for a new faculty member. They may have similar ages, mathematical interests, or other interests (e.g. young children), and it is natural for them to socialize with each other, to offer advice, and, in general, to make themselves available as trusted listeners who are "on their side." Similarly, departments with well-defined research groups may have senior faculty in a research group for whom it is natural to serve as mentors for young faculty members in the same group.

Whether or not there are natural mentoring relationships developing, the chair should be alert to the needs of each untenured faculty member and should try to put solutions in place even before problems arise. As a department chair, I tried both formal and informal mentoring relationships. For example, I encouraged new faculty to participate in Project NExT, because I believe it helps build an important professional community for new faculty and because the program provides valuable mentoring for new members of our profession. I also tried to make certain that new faculty always had someone to turn to for information and support. At important points in the journey from new hire to member of the tenured faculty, I would seek out senior faculty members and ask that they make themselves available to particular faculty both to offer good advice and to be an advocate for that junior faculty member.

In addition, as chair, I tried to be available to all untenured faculty and to communicate my interest in being both helpful to, and supportive of, untenured faculty. Periodically, I would meet with untenured faculty, alone and in groups, to make certain that they understood university procedures related to annual evaluation, tenure and promotion. I also had an open door policy at the office. If an untenured faculty member dropped by my office and asked, "Do you have a minute?" I frequently responded "Come in, I always have time for you" no matter how busy I might have been. My goal was to communicate to the faculty member that I wanted him or her to be a success; I would give willingly of my time and effort to help the faculty member be a success; and that I believed he or she would be successful.

Different faculty had very different needs. For one faculty member, we talked at great length about teaching and how he could develop a better rapport with his students. Another wanted detailed information and an extended discussion of his progress towards earning tenure. A young woman on our faculty who was getting too many opportunities for service roles on campus needed my encouragement in order to say "No" to some of the invitations. Some needed to be told that they needed to focus more of their energy on their research program. Some needed reassurance that they were doing enough—in fact that the senior faculty were thrilled with their achievements. Others needed to be told that it was OK to relax and that quite possibly they were working too hard.

The department chair's responsibilities as a campus administrator will eventually be in conflict with the goal of being a good mentor for untenured faculty. The department chair has the lead responsibility for reappointments, for leading a promotion and tenure review, and for making a recommendation for or against a faculty member's tenure or promotion. For example, whenever a faculty member was reviewed for tenure or promotion, I would meet with him or her and carefully go over our campus' procedures. This included the process by which we seek external reviews of the faculty member's research. It was my job to explain the candidate's right to review all letters that are received or, at the opposite extreme, to waive their right to know who was asked to review their work and their right to see the letters received. I could not offer any advice as to what the candidate should do because it is important that a decision about what to do be made without any pressure to make a particular decision. At this point in the process, I urge the candidate to seek the advice of a senior faculty member they trust, and I urge the most appropriate faculty member to accept this role if asked.

One of the keys to a good mentor-mentee relationship is that there is a solid basis for the relationship, whether it develops naturally or is organized formally. My own experiences as a mentor have varied greatly. I have participated in a campus-wide mentoring program with, at best, modest results. While I was willing to be a mentor and wanted to be supportive of the program, there was no foundation on which the relationship was built. On the other hand, when my dean once asked me to mentor an outstanding young faculty member who had had a major falling out with his chair, I think I did a better job. I knew why I had accepted the role and I believed that my experiences made me a good choice for the role. In another situation, I was highly motivated to be a good mentor for a young colleague in another department with whom I shared research and teaching interests.

Within our own department I was, of course, highly motivated to help our young faculty succeed. Indeed, I consider it close to a chair's most important duty. After all, if you help a faculty member develop professionally to the point where they are very successful at their job, you add value to the investment the Department made when the person was hired, and you make the tenure and promotion process much easier for both you and the faculty member. In my approach to establishing good senior faculty–junior faculty mentoring relationships, spreading out the workload was never a consideration. Instead, I always sought a senior faculty member who would have a natural interest in the success of the faculty member, and I looked for faculty I expected to be good as mentors. (I recognize that this may come across as another example of the cliché, "No good deed goes unpunished.")

If your department does not have a well-established mentoring program, you may be concerned with how a good mentoring program is initiated. To state the obvious, mentoring can be either informal or part of an organized mentoring program. Also, mentoring programs can be professional (like Project NExT), campus-wide, or departmental. Different programs have different purposes and the existence of one kind will not always eliminate the need for a different kind of mentoring. As this essay is directed primarily towards department chairs (or more generally towards the senior faculty in a department), I will repeat that I see it as the chair's duty to be certain that programs are in place that meet the needs of the department by meeting the needs of the junior faculty, whether that program is informal or formal.

Having said this, I would encourage any untenured faculty member not to wait for a mentoring program to be established when none exists. Recommend that they pick out someone they respect (whether it is the chair or another experienced faculty member) and ask for his or her advice and help. Quite likely, the faculty member will be eager to help, and flattered to be asked, even if in need of some guidance as to how best to offer their assistance.

Dependence On and Culturalization of Part-time and Temporary Faculty

Norma M. Agras
Miami-Dade Community College
Wolfson Campus

Over the five years during which I have served as department chairperson, student enrollment in my department has grown 67%. Yet though we were already understaffed in 1998, we have only hired one additional full time faculty member, for a total of 12, in a department that served 4,500 students this spring. Thus, the stage is set for an ever-increasing dependency on adjunct faculty.

This past spring adjunct faculty taught well over 50% of our credits, teaching mostly developmental classes (below the level of College Algebra). In some cases adjuncts also taught some of our most advanced offerings. The adjuncts who work in our department are for the most part industrious, caring instructors. Yet as often happens in other departments and perhaps at other colleges, they are in some ways treated as "second-class citizens." I find this to be disturbing at best, and have taken various steps to be supportive of them and encourage their commitment to our program.

Adjuncts receive from me information about college events and about workshops facilitated by the college (e.g., Excel or collaborative learning) or by our department (e.g., teaching with graphing calculators or other technology). I send them any information that I receive about full time teaching opportunities. I also send them information about the MAA, AMATYC (American Mathematical Association of Two Year Colleges), and other professional mathematics organizations.

Adjuncts are expected to communicate with their students. This is done via email (each adjunct has an email account) and via a telephone extension where students can leave them messages. Adjuncts are required to hold office hours in our mathematics resource area that we call our Math Lab. They receive information about department and college policy, and receive sample course handouts. Those who teach college algebra or above are given a graphing calculator with an overhead panel.

Adjuncts are invited to department functions such as luncheons or holiday parties and some department and college-wide mathematics meetings. I ask for their input on textbooks or course software. I try to be as accessible as possible and often ask them how they are doing, how their classes are going, etc. If I don't see them, I email them. I feel strongly that a chairperson needs to keep open lines of communication with all faculty members.

Several faculty members do not share my desire to include adjunct faculty in our department activities, nor do they approve of my willingness to listen to adjuncts' ideas. As one faculty member put it, "They are JUST ADJUNCTS!" This attitude led to the most significant conflict I have had in five years with my faculty. There was a confrontation in a spring department meeting, and our department has never completely healed from the rift. Luckily, one of the faculty members who was most resentful of the adjuncts has transferred to another campus. With several people retiring, my department will be significantly transformed this fall.

As a department chairperson, you have to walk a very fine line. You must support the full time faculty and the department culture. At the same time you must support the adjuncts and involve them to some degree in the department for the sake of a more cohesive program that better serves the students' needs. Regardless of the conflict and its effect on my department, I continue to believe that adjuncts need to feel they have the support of the department chairperson and that their contributions are valuable to the department.

Another way to support adjuncts is to have full-time faculty mentor them. I attempted this with my department, but this kind of mentoring was quickly rejected. I hope to attempt it again next fall. Another strategy that could improve the teaching performance and expectations of adjuncts is to write performance reviews for them even if under different, less stringent guidelines than those of full-time faculty. Additionally, a reward mechanism could be created with awards such as Adjunct of the Year. Multiple-level or tiered adjunct contracts might also be effective in recognizing and rewarding adjuncts for their contributions to the department.

Dependence On and Culturalization of Part-time and Temporary Faculty

Catherine M. Murphy
Purdue University, Calumet

Without the availability and support of our non-tenure track teaching staff, fewer than half the sections of our courses could be taught. This staff consists of four Continuing Lecturers (permanent, non-tenure track), four Visiting Instructors (full-time temporary), usually four to six TAs, and twenty-five to thirty five limited term lecturers (part-time). None of these categories have faculty status.

Despite the university's definition of continuing lecturers as "not faculty," the department welcomes them as full colleagues and treats them as experts in freshman/sophomore level service courses. They serve on appropriate curriculum oversight committees, have input on assessment issues, and are expected to be involved in professional development. The department supports those activities as we would for tenure-track faculty.

Visiting instructors are expected to have contracts for two consecutive years. No more than three consecutive years are allowed. They are invited to participate in departmental life as much as they wish. We've been fortunate that we've been able to hire visiting instructors who are very good classroom teachers. We usually have a couple who also want the experience that involvement in departmental service and activities give them.

Our TAs are Master's students in mathematics, engineering, education, or management. We require them to attend pre-semester teaching workshops, hold office hours within the department, and meet with their course coordinator on a regular basis. These TAs spend a lot of time around the department and are invited to department meetings and social occasions.

Limited term lecturers (LTL) are allowed to teach at most two courses per semester. Most of them teach at two or more different institutions. We require pre-semester teaching workshops (one day's duration). For certain courses we've also started requiring attendance at weekly meetings with the course coordinators. These are usually scheduled in late afternoon; pizza is served. Experienced LTLs are allowed to miss up to half of the meetings without jeopardizing future assignments. So far most of the LTLs involved in this collection of courses have found the experience valuable and are willing to accept this requirement as part of the job description. We are considering expanding the use of LTLs to more courses. LTLs are invited to attend department meetings and we regularly survey their opinions of text books, syllabi, pace of courses, etc. The schedules of most LTLs preclude very much involvement in the department.

The department's goal is and has been to reduce our reliance on part-time instructors to at most 25% of our offerings. The creation of the Continuing Lecturer classification and permission to hire full-time Visiting Instructors has helped substantially. However, part-timers still teach about 35% of our offerings in Fall and 30% in Spring. The department has hired all the Continuing Lecturers it can under the university's guidelines. Although this will make further reducing our reliance on part-time instructors more difficult to accomplish, the department's responsibilities in the areas scholarship (faculty, graduate, and undergraduate), as well as university and professional service, require at least maintaining the size of the tenure-track/tenured faculty.

Dependence On and Culturalization of Part-time and Temporary Faculty

Donna Beers
Simmons College

Culturalization of Part-Time Faculty

The reliance on part-time faculty by colleges and universities has increased steadily over the last thirty years. "It is widely known that the proportion of all faculty who teach part-time virtually doubled from 22 percent in 1970 to 43 percent in 1997 (National Center for Education Statistics 2001)." [6] According to statistics from the U.S. Department of Education, published in the Chronicle of Higher Education, [7] the percentage of faculty employees was 35% in 1983, 40% in 1993, and 43% in 1999. "Today, 43 percent of all faculty are part-time, and non-tenure-track positions of all types account for more than half of all faculty appointments in American higher education." [2] Given these facts, it is urgent that departments work to ensure that their part-time faculty are highly motivated and well prepared to carry out their teaching duties in the department.

However, part-time faculty often live at the margins of departmental life, only present on campus to teach their classes. The economic reality of low pay makes it urgent for many part-time faculty to 'hit the road' in order to get to their next teaching assignment. Because they are not voting members of the faculty, they are largely unaware of the departmental or school culture in which they are teaching.

Where do the Problems Lie?

A range of obstacles prevent part-time faculty from becoming acculturated to an institution. Financial pressures are a major obstacle, including low pay, no benefits, and expensive parking fees.

Also, part-time faculty often feel like second-class citizens. They have no voting privileges. They sometimes are treated as unworthy by tenured faculty. They receive no committee assignments. They are assigned the least attractive offices or have to share an office with several others. They are assigned the least attractive teaching schedules. They often do not receive basic administrative support because they teach late afternoon or evening courses.

Many part-time faculty feel exploited. In addition to earning low pay, they teach large, introductory classes which the full-time faculty don't want to teach. Also, they feel they carry heavier workloads than full-time faculty because they teach large service courses with potentially weaker students and larger grading loads.

Strategies, Warnings, and Opportunities

Department chairs must take the lead in helping part-time faculty to become productive and engaged members of the college community.

First, to motivate part-time faculty and to keep them from feeling exploited, chairs need to treat part-time colleagues fairly and respectfully. This means providing them (in a timely manner) with office space equipped with the usual appurtenances afforded full-time faculty, such as office furniture, phones, computers, keys to their offices, e-mail accounts, voicemail accounts, and course management accounts. It also means ensuring that part-timers have access to basic administrative support such as photocopying and helpdesk support. Chairs should use all available means to secure parking fees for their part-time faculty that are proportional to their salaries (or even waived, if possible). They should make their part-time faculty aware of any in-house faculty training opportunities or pedagogical workshops. They should also mentor part-time faculty through classroom visits (although no more than any other faculty) and through feedback from student evaluations.

Second, to maintain ongoing communications with part-time faculty in order to make sure that things are going well, chairs need to be accessible. An open-door policy may encourage faculty to come in and discuss any problems or questions. Regular meeting times are preferable to hallway conversations that take place 'on the fly.'

Third, to help part-time faculty feel connected to the department and to the institution, chairs can foster a sense of belonging by inviting them to department social events, soliciting their input regarding issues of departmental concern, affording them public recognition for their teaching excellence and/or their particular expertise, providing them opportunities to teach new courses they have been wanting to teach when possible, and providing them financial support to attend a professional meeting or workshop that is relevant to their teaching assignment at their institutions, whenever possible.

Warning: 43% today will quickly rise to 50% and higher. It may be time to consider the issues awaiting full-time faculty once the part-time faculty are the majority. What is the likely reaction of parents once they realize that their tuition dollars are being spent for a faculty who are not committed to their college on a full-time basis?

Dependence On and Culturalization of Temporary Faculty

Connie Campbell
Millsaps College

At Millsaps, there are two basic types of temporary faculty—fixed-term and extended term. Fixed-term faculty are those who are hired specifically under a short-term contract with the general understanding that it is not likely that their position will be renewed beyond the initial term (i.e. sabbatical replacement); whereas extended term faculty are non-tenure track faculty who are basically assured that, given adequate performance, they will be rehired on an ongoing basis until the program undergoes a substantial change.

The issues of dealing with fixed term positions are very similar to that of adjunct faculty. While these faculty typically are provided with office space and are more engaged with members of the department, there is a real temptation on the part of the chair to avoid conflict by not offering direct feedback. Additionally, there is typically no real commitment on the part of the institution to help these individuals develop as teacher/scholars. Nevertheless, the chair should find ways to support, encourage, and motivate these faculty.

Millsaps College currently employs two non-tenure track mathematics faculty. Each of these individuals holds a Master's degree in mathematics and has been teaching full-time for the department for more than ten years. These positions are generally understood to be continually renewed, even though the contracts are given on an annual basis. The primary teaching load for these faculty is the same as for tenure-track faculty. However, they typically only teach courses which are at or below the level of Calculus II. Non-tenure track faculty are expected to fully engage in areas of scholarship and service and are evaluated annually under the same criteria as tenure-track faculty. This produces some level of frustration among the non-tenure track faculty, particularly since it is extremely difficult for them to fare well in comparison to their peers in the area of scholarship. However, the system does establish and reinforce the college's commitment to the continual professional growth of all faculty members. While these faculty may struggle to find success in the area of scholarship, they have proven themselves to be vital assets to the department as outstanding teachers and major contributors to the life of the college.

One common problem with non-tenure track, extended term faculty is that they often feel like second class citizens. Even though they are paid less, they are expected to perform at the same level, and are evaluated by the same standards, as tenure-track faculty. Raises are given terms of a percentage increase, and non-tenure track faculty are not eligible for promotions. Given that it will be extremely difficult for them to perform well in the area of scholarship, which represents one-third of their annual review, and that they are not in a position to go up for promotions, it is not realistic to expect that their salaries will ever increase by a substantial amount.

It is the job of the chair to encourage these faculty, help them see they make important contributions, and offer feedback and guidance. Good teachers are hard to find and the frustrations that these individuals have are valid. As their department chair, it is my goal to find ways to help these individuals develop in the area of scholarship and also to find ways to use their strengths to enhance the department. Outstanding teachers are an asset to the department and as chair I want to make sure that these individuals know that they are valued and that they are substantial contributors to the department and its mission.

Resolving Conflict Within the Department

Norma M. Agras
*Miami-Dade Community College
Wolfson Campus*

In addition to conflicts between faculty and me, which I have tried to avert whenever possible over these past five years, I have been faced with several conflicts among faculty and among members of my staff.

Some conflicts have been professional in nature, in some cases with professional jealousy deeply rooted but obscured. There have been disagreements on how instructors should teach, including the use or non-use of technology and vehement disagreements with regard to textbook selection. More serious problems have been caused by some faculty members taking credit for others' work or ideas.

Conflicts have been as varied as the personalities in my department. However, there are common threads in how I have been able to help resolve these conflicts. The most important thing that a department chairperson can do is to listen. How you listen is as important as, or more important than, the act of listening itself. You need to listen dispassionately as each party expresses his or her beliefs or feelings on the subject that is at the root of the disagreement. Regardless of how quickly you think you have sized up the situation, both parties must feel that you are impartial and that you will not draw conclusions or make recommendations until after both parties have had a chance to speak.

Although it could perhaps be said that war has resolved many human conflicts, war has no place in your department. Once you have listened to both parties and have had a chance to think about both sides of the issue, it is best to bring both parties together in a place where there will be no interruptions so that a dialog can take place. It is at that time that you might need to make a call based on information from both parties. It is important that you stick with your decision, and that you do not personalize the problem. Both parties need to feel that you are supportive of them regardless of who in your final opinion was correct in some conflict or situation.

In a case in which one person took credit for another's work, I needed to step in and take immediate measures to stop the situation before it escalated into a legal mess.

Above all, the department chairperson must be a peacemaker, an impartial judge, a mediator between any two conflicting parties. Human conflict is evident everywhere. How you handle it could have very beneficial or disastrous effects on your department.

Resolving Conflict Within the Department

Connie Campbell
Millsaps College

Conflicts within the department will most certainly occur, and it is up to the chair to find ways to make sure that disagreements among faculty members do not impact the department and its program. Chairs should work to foster good group dynamics, helping department members to see that each person has the capacity to contribute to the overall vision of the department in some unique way. While it is not necessary that members of the department be friends, it is necessary that each faculty member maintains working relationships with others in the department.

Not surprisingly, many chairs prefer to avoid conflict, hoping it will resolve on its own, only to find that situations often worsen over time. For those with this tendency, as well as those who meet conflict head-on, the following list of tips may prove helpful in preventing conflicts within the department, thereby minimizing the time and energy spent addressing unnecessary problems:

- Treat students, staff, and faculty with respect.
- Don't ask anyone to do something that you would not be willing to do yourself.
- Communicate your ideas and ask for input.
- Discuss changes you would like to see made prior to executing those changes.
- Identify potentially volatile areas (i.e., scheduling, departmental spending), request input, and provide clear communication of relevant policies, practices, and expectations.
- Establish a policy about decision making and be consistent with it.
- Make it clear that disagreements among faculty are not to be discussed in front of students, as this is highly unprofessional and only compounds problems.
- Validate faculty members' different points of view on a regular basis.
- Listen, listen, listen.

When dealing with conflict, chairs should be careful not to come to conclusions before having all the details. Almost always there is more to the story than first appears; consequently, the chair should make sure all voices are heard and facilitate a fair and thorough process for conflict resolution. Particularly when the conflict involves programmatic issues, having such a process can keep the issues on a more professional and objective level, enabling a more comprehensive assessment of the program.

It is the job of the chair to try to bring out the strengths in everyone and then to try to create an environment where every member of the department makes a contribution. The chair should work to try to find a way to use each person's strengths for the betterment of the department. Also, try to avoid unnecessary conflicts. If two people do not get along, they should not be put together on a committee of two. The chair's challenge here is to find a way for each member to contribute to the department in such a way that the whole is better. The result is that while faculty may not all like each other, each member feels valued and has a voice in the overall program.

We are living in a time when documentation is imperative; however the chair should be discrete when choosing the appropriate medium for discussions. When problems are raised, the chair should communicate promptly and effectively with faculty. The chair might choose any of the following means to engage the issue: personal conversation in a public space, personal conversation in the faculty member's space, personal conversation in the department chair's space, e-mail dialogue, or formal letter. As listed, these range from less intrusive to more serious modes, and care should be taken to make sure that the mode of communication is appropriate for the discussion.

E-mail and letter responses are more formal and may lead to misunderstandings due to tone. Furthermore, such media are utilized for documentation and, consequently, may convey to the faculty member that the issue is more serious than it really is. For example, if someone has spoken with you about an unofficial complaint and you want to discuss it with the faculty member, e-mail would not be an appropriate medium. However, if the issue involves an official complaint, then the chair should document that he or she has addressed the issue with the faculty member. In such instances the chair might begin by e-mailing the faculty member to request a meeting, providing some background for the meeting and, following the meeting, provide a written summary of the meeting to each person involved.

Managing Conflict with the Dean, Provost and Other Departments

Connie Campbell
Millsaps College

Dean/Provost

It is critical that the department chair have a good working relationship with the academic and Division deans and, toward this end, chairs should be very careful to pick their battles. While their supervisors will certainly make some decisions with which the department disagrees, some things are clearly not worth arguing over. However, when conflict does arise, it can be resolved much more easily if the chair has laid a solid foundation from which to address the issues. The following are some general guidelines for chairs in building such a foundation:

- Know the supervisors' expectations and work toward them.
- Demonstrate respect for supervisors to department faculty.
- Meet deadlines.
- Stay within budget.
- Build support for ideas from the bottom up.
- Avoid going over the supervisor's head.
- Listen effectively.
- Communicate any potential problems of which the supervisor should be made aware in a timely manner.
- Seek to understand the big picture.
- Help administrators understand how your requests/actions are in support of the larger goals of the college.

Such actions will set the stage so that when conflicts arise (and they will), the dean will have a positive overall attitude toward the department chair, knowing that the chair has the department's and the college's best interest in mind.

Additionally, it is crucial that the department chair find a way to communicate effectively with his or her supervisors. Indeed, communication is key. The department chair must have a clear understanding of the administrator's expectations so that he or she can effectively meet and/or shape them. On the same note, the chair needs to make the department's expectations and needs clear to the administration. It is also helpful if the chair knows the management style of the supervisor (i.e., does the supervisor want comprehensive documentation, or just the highlights; does he or she want to be kept informed of little things that may become big things or would he or she prefer that you handle the little issues and only give the background information if a larger problem arises; does he or she respond better to e-mail or face to face?)

Other Departments

A good working relationship and sense of mutual respect between the mathematics department and other departments are critical. Mathematics faculty should work to make sure that they are not perceived as a separate, elite entity. An excellent vehicle to facilitate communication between the mathematics department and other departments would be a commitment to working with those who have a mathematics requirement as part of their major. The mathematics department should recognize that these services to other departments constitute a significant portion of its course load and make sure that the needs of the other department are being met, while also maintaining the integrity of the mathematics curriculum. To this end, the department chair should periodically discuss the curriculum of these courses with the chairs of the various departments, including course materials and the topics which are covered. Additionally, the department chair should work with all of its constituents concerning the scheduling of required courses to minimize schedule conflicts with other required courses. It should be noted that effective communication with other departments involves not only understanding their needs but also making the mathematics department's needs known. Such communication goes a long way in earning the respect of other departments as well as preventing unnecessary problems. Moreover, when the mathematics faculty strive to broaden their vision to incorporate that of other departments, hopefully the endeavor will be reciprocated, thereby providing students with a more comprehensive educational environment and faculty with a more cooperative one.

Another way to build relationships among various departments is through service on college-wide committees and other more global activities. In addition to building relationships with other departments, these types of activities also help those outside the department see the mathematics department and its faculty as an integral part of the institution.

Managing Conflict with the Dean, Provost and Other Departments

Jimmy L. Solomon
Georgia Southern University

If a conflict arises with the dean, then it must be resolved. To a lesser degree this is also true of the provost. However, if you and the dean work well together, then your conflict with the provost can probably exist without a complete resolution.

Any conflict with the dean must not result in your inability to adequately represent the department. If this can not be achieved, then your tenure as department chair will probably not continue for long. If you have tried with no success to resolve your conflict with the dean and you are convinced that a resolution must be found, then you should consider approaching the provost, although you can probably be assured that the situation has been discussed between the dean and the provost. Before approaching the provost, you must have confidence that the causes of the conflict are overwhelmingly the fault of the dean.

In my basic effort to interact with the dean or provost, I try to determine answers to the following broad questions:

- Does the individual have a clear understanding of what he or she wants in a department chair? If so, are you comfortable with that expectation?
- Does the individual like verbosity or succinctness?
- Does the individual like a formal or informal atmosphere in discussions?
- Does the individual like to see hard data?
- Is the individual committed to the role that the department is playing in the college or university?

One definite no-no: on rare occasions, in order for you to properly represent your unit, you will have to challenge the dean on his or her response to something which you think is absolutely in the best interest of your department. However, you can never do this in an open situation. If you attempt this, you can probably expect not to continue as chair for an extended period—always show respect for the position of dean (provost), and maintain "public" respect for the individual occupying the position.

The good news is that severe conflicts between a chair and dean (provost) are relatively rare, but when they do surface, have confidence that a resolution to the conflict will be found.

To minimize conflicts with other units, you have to maintain a line of communication. If you have a large "service client," for example, the College of Engineering or the College of Business, you are well served to have a committee with representation from the service client and the department to monitor courses and other joint efforts. Any serious conflicts between departments usually find their way to the provost.

Finances

External Funding
Donna Beers
Martha Siegel

Budgets
Catherine Murphy, Jimmy Solomon, and Tina Straley

External Funding

Donna Beers
Simmons College

Deans and academic vice presidents regularly encourage departments to secure external funding for a variety of initiatives. These include outreach projects, programs to enhance teaching, interdisciplinary research, support for undergraduate research programs, equipment, and summer research by faculty. For mathematics, there are several sources of external funding, both public (e.g., National Science Foundation, National Institutes of Health, National Security Agency, U.S. Department of Education) and private (e.g., ExxonMobil, Akamai, Tensor, and Sloan Foundations).

We focus here on external funding issues surrounding proposals for collaborative programs such as outreach programs and interdisciplinary programs.

Pros and Cons of Securing External Funding

There are many positive reasons for department chairs to encourage and lead grant-writing activities to secure external funding. Altruism, or the sense that 'it's the right thing to do,' offers a strong rationale for pursuing external funding. Proposals envision the creation or delivery of products or services that are needed by target audiences identified in the grant proposals.

In addition, successful grant writing offers significant benefits to institutions and departments: It enhances visibility and reputation; it provides resources (for example, equipment and stipends for faculty and students); and it provides faculty with experience in grant-writing, leadership, and project management, which contributes to their professional development. Publicity from external funding activities involving students may also help to recruit new majors.

On the other hand, there are some disincentives to pursuing external funding. First, externally funded programs that take place on the college campus will compete with regular courses for classroom and lab space. This can present significant scheduling problems and requires ongoing communication and coordination with the Registrar's office to confirm commitment of facilities.

Second, externally funded programs require personnel resources (college faculty and/or students and/or support staff). These personnel needs compete with departmental instructional needs and existing institutional programs (e.g., mathematics majors are tapped to participate in programs such as Upward Bound, America Counts, etc.). These needs must be carefully and realistically weighed before a department or institution commits to engaging in new funded programs.

Third, there are potential risks to faculty who engage in externally funded outreach or other collaborative activities. The absence of clearly stated incentives or rewards for participation in externally funded programs, either in annual performance review criteria or in institutional tenure and promotion guidelines, may jeopardize a faculty member's chances for tenure or promotion. In addition, inadequate release time or inadequate administrative support may lead a faculty member to divert time and energy from other professional activities, which also may jeopardize tenure or promotion and/or merit pay increases.

Finally, there is considerable overhead in maintaining smooth communications and coordination with partners (both internal and external) on externally funded projects. Partners necessarily have their own agendas and priorities.

Where do the Problems Lie?

With externally funded collaborative programs, there are challenges at every stage. The initial challenges are to establish a clear rationale for the project, identify the target audience, identify key players, develop shared goals and objectives, and develop a timeline and budget. Project planning is crucially important, but finding a regular meeting time for ongoing review of goals and progress toward achieving them can be problematic.

Numerous operational details need constant attention, e.g., reserving multimedia equipment and facilities, purchase of equipment and software, placing catering orders, payment of student and faculty participants, and working with the office of public affairs to publicize program events. For a faculty member running a program for the first time, juggling these details along with teaching and scholarship can present time pressures.

Another challenge is to measure the quality of the program. Assessment is an important and required dimension of any program. Sufficient time must be devoted to developing and administering pre- and post-assessment tools, evaluating the results of the assessments, and securing external reviewers.

Other problems in carrying out an externally funded program include: establishing clear lines of communication (who reports to whom) and identifying what information is distributed to whom, from whom, and according to what timetable; managing personnel, including setting expectations, motivating participants throughout to perform their assigned roles, providing participants feedback so they can assess their success in carrying out their roles and be accountable; and identifying and managing risks.

Advice on Strategies, Warnings, and Opportunities

Before departments or faculty embark on externally funded activities, they need to confirm that these activities are an institutional priority. They also must identify an administrator who will actively and publicly champion the endeavor, provide guidance, communicate to all relevant offices that this endeavor is an institutional priority, and set expectations for their cooperation.

Faculty who are experienced in obtaining and administering external funding in a similar project should mentor faculty applying for a grant for the first time.

Success with—and lessons learned from—small pilot projects can give departments the experience and confidence to design larger scale projects, to pursue larger externally funded opportunities, and to make grant-writing a habit of mind and regular activity within the academic year.

External Funding

Martha Siegel
Towson University

External funding to meet the needs of a department, individual faculty, or students is desirable and should be sought and considered a regular part of a department's workload. Once a proposal is funded, the real work begins. Every department needs to have competent help in preparing proposals and serious support in implementing the plans should it get funding. All potential expenses and efforts should be recognized before submitting the budget. In order to be successful, a grant proposal should be focused, well developed but limited in scope to the task at hand, and follow the RFP (Request for Proposals) carefully.

Grants to Support the Curriculum or Teaching

The department chair should be sure that the department agrees with the goals of the proposal. For example, an individual faculty member wants to apply for funds to reform the freshman college algebra course, but 80% of the faculty, whom we might call the more traditional thinkers, are not interested in any such change. If and how should this proposal go forward? What should be the scope of the implementation plan in the proposal?

Faculty members working on the project will need some reassigned time. Does the grant adequately cover the costs of faculty replacement? In fact, can an essential faculty member be replaced on an interim or part-time basis? As an example, consider reform of teacher education where the faculty member teaching mathematics education courses gets reassigned time. If that faculty member is the only one, or one of the few teaching mathematics education courses, who covers her classes?

There are several types of outside support that seem to benefit a comprehensive Master's-degree granting institution. First, there are grants that help departments purchase equipment and laboratory enhancements so that the technology needed for effective teaching can be available in the classrooms. These must be sought with a clear educational focus in mind—curricular enhancement and development of new courses to meet true educational needs. Since these generally require matching funds from the institution, the school has to have a commitment to the project from the beginning. Such projects require that professional development for faculty will be available so that they make good use of the equipment in their teaching of mathematics.

Grants to Support Students

Applying for grants to support students, such as NSF's VIGRE (Vertical Integration of Research and Education in the Mathematical Sciences) and CSEMS (Computer Science, Engineering, and Mathematics Scholarships), may be the sort of thing that the chair has to initiate. Application for an REU (Research Experiences for Undergraduates) must be initiated by particular faculty. How can the chair encourage and facilitate this type of inquiry-based learning and help to make this part of the culture of the department? What will be the impact upon the department? How should such an impact be evaluated before an REU proposal is submitted? Do these programs require another person as coordinator or principal investigator? Is there administrative support for undergraduate research? Will there be adequate recognition and workload adjustment for those who serve as mentors?

Are there other programs to which a department may apply for more support for students? One or two programs within NSF allow departments to apply for scholarship funds for undergraduates. State resources and private funds may also be available. The Noyce Scholarships are available to fund undergraduates who are majoring in STEM (Science, Technology. Engineering and Mathematics) but agree to get degrees in secondary education in these areas of need and to teach in secondary schools in locations designated as high need areas for at least two years. The funding for individual students need not be need-based, but need could be a factor in making the local decision on any given student. These funds are also available as one-year stipends to graduate students who have degrees in STEM and take an education program that will give them certification.

The CSEMS program at NSF does require that students have some level of need and are majoring in computer science, engineering or mathematics. The money for this program comes from fees that companies submit to the USCIS (U.S. Citizenship and Immigration Services, formerly the Immigration and Naturalization Service (INS)) in order to employ foreign nationals when American technical workers are not available in the labor pool. Many institutions are not aware of these programs.

Grants to Support Individual Faculty Research

Faculty research and travel grants are available from many sources. The department chair may have to prioritize when several regular faculty members wish to pursue their research careers with supported assigned time during the same semester.

Support for Sabbaticals

The department chair should encourage faculty to take one-year supported sabbaticals. Most institutions will provide 50% salary to those on sabbatical for one year (and full salary for those on sabbatical for a semester). Having faculty on leave for a year saves the institution money and may allow the faculty member to take on a more serious research project. Chairs should try to facilitate such support for good research projects. These days, with budgets what they are, many schools will be far less accommodating on sabbaticals. What should a chair require of a sabbatical proposal to feel comfortable giving it his or her approval?

Assessment and Evaluation

It is important to remember the assessment piece of any proposal for outside funding. Chairs should encourage and lead the proposal writers to provide a strong assessment and evaluation plan.

Budgets

Catherine Murphy
Purdue University, Calumet

Jimmy L. Solomon
Georgia Southern University

Tina H. Straley
Mathematical Association of America

Often the success or failure of a department chair will hinge on the ability to manage, and increase, the department's budget.

There are two basic musts for any chair in order to maximize the departmental budget:

- Develop a complete understanding of the budget process on your campus, particularly as it pertains to your unit.
- Keep your own financial records at the unit level—independent of the central accounting office. Do not rely on the central accounting office to keep you abreast of your finances.

A department budget has two main parts: salaries and operating expenses. Although the personnel or salary portion of the budget is the largest part, you have less decision-making power over it, except to argue for increases. The discretionary part of the budget, the operating budget, is relatively small. But how you decide to spend it is of vital importance. There are items that may or may not be under the discretionary powers of the chair such as student assistants, TAs, or adjunct faculty. The ideas below describe some general strategies and need to be adapted for your department.

Having more money means being able to hire more faculty and staff, being able to give more course release time for research and special projects, supporting travel, buying new equipment, and providing other faculty and staff development. To increase your budget: make a very good case for the increase; be persistent and consistent in your requests; and show that the increase will benefit the entire college, school, and university. Present your arguments convincingly and often to the dean, the provost, and anyone else who will listen and who can help.

The budget process differs from institution to institution and the level of decision-making the chair has also varies considerably. The best advice is to know how it works at your institution. Find out who is successful and learn from them. Getting perks from the administration is a mark of a successful chair. Again, you are the advocate for your department, faculty, staff, and students. It is your job to see they get all they fairly can and to which they are entitled. Given these caveats, there are approaches that work well from campus to campus.

Personnel Budgets

Personnel budgets include full-time faculty, graduate assistants, part-time or adjunct faculty, student assistants, and staff. Make sure that your department personnel are being treated fairly in comparison to other departments in your institution and in similar institutions. Make sure the number of lines you have are competitive. Not getting your due in either numbers of positions or in salary levels is an issue that you should not let go. If your levels of support are consistent with other departments, getting more personnel lines in any of the categories must be based upon greater need. You have to make a very convincing case, and you will likely have to repeat the request and continue to make your case.

Faculty release time, special projects, and growth in programs or enrollment are among the reasons for increasing personnel. If you have release time paid for by grants, you will most likely replace the grant-supported faculty member with TAs or part-timers whose salaries are much lower than the faculty. Make sure that your department retains control over the difference, which is money saved by the college that can be redirected to new purposes. With it, you can hire additional part-timers or pay TAs in addition to the one-for-one replacements. You can give the grant team additional released time or give members of the department released time for other, non-funded work. The latter approach spreads the benefit of the grant-supported project through the department and is a good way to get the support of the other faculty for the work being done. If projects are being done for the administration, you should request an appropriation to support the work, for example to hire faculty to cover classes. If during the budget process for the following year, you can estimate the income that your department will receive for faculty releases, you might be able to pool the money to hire full-time faculty rather than part-time.

Unless your institution is unionized, one way to increase personnel budgets is to request high merit raises for your faculty. Don't be bashful in your requests. Remember that you are the only one fighting for the mathematical sciences on your campus. Request as much as you can. If there is a rating scale, start your best people at the top of the scale. But don't overreach. If you request the top raise for all of your faculty, then you have diluted your argument. The conclusion might be that all of the faculty in the department are on the same level, hence all are average. The same argument applies to staff. While you are department chair, you have to make the hard decisions and you must differentiate between your faculty and staff in requesting raises. To do so, you must have good review policies and procedures and you must adhere to them and do your part thoroughly and conscientiously.

Operating Expenses

The operating expense budget, which comes under different names, is the area over which you have most discretion. Of course you have to buy supplies and there may be other areas for which you are charged and have little control, such as telecommunications. However, there are areas over which you may have total control such as equipment and travel. The morale of faculty, students, and staff is greatly affected by the relatively small operating expenditures within the total department budget. Perhaps because the use of discretionary money is so highly coveted, some department chairs and other administrators like to keep details of the budget a secret. As the saying goes, "knowledge is power," but it is really only powerful when it is shared. Bringing your department faculty in on the facts and soliciting assistance from them on how to dole out the money, especially for travel and equipment, are the best ways to gain their support for those tough decisions you have to make. In the end, however, the budget is your responsibility and you have to make the decisions.

Operating expense budgets can be a source of frustration for a chair because they are always below the levels needed. Some may increase by an inflationary amount across the board but will not reflect real increases that result from other pressures. For example, mathematics departments had very small equipment and supply budgets when all they needed was blackboards and chalk. The need for computers in the classroom may still not be built into many department budgets. Photocopying charges have grown rapidly and can quickly exceed your resources. You might have to keep reins on this expense by enforcing user accounts and codes and discouraging making class sets of notes or copies from books. Use of long-distance telephone calls is something that people might not think twice about, but it can be a place where department expenses can be reduced. Library budgets are often areas that get cut when money is tight and then are not adequately restored. Recently, some journal subscriptions from commercial publishers have increased radically and libraries are cutting these subscriptions. Choosing among these expenses are hard decisions that you might have to make. Since the department faculty do not want cuts anywhere, you are wise to share the constraints with them and allow them to have input on what is most important to fund.

Even though the operating budget is set for the year, you often have the ability to move the money within it. Paying attention to these details and keeping your budget and expenses in order will be an advantage when you have to submit the next year's budget.

Sources of Additional Funding

Although it is true that the dean often has discretionary money that you can request, this is generally not a continuing source of funds to augment your department budget. Instead, you have to find ways to bring more money into the department on a more reliable basis. You should know your institution's priorities and strategic goals. If you can fit department wishes into the strategic plan, there may be money and other resources that will come your way. Consider collaborating with another department that has supported research or other projects. Some resources may directly benefit your department or may free up money that you can redirect. Work with the institution's development office. If donors target your department, not only will you have more money to spend but this money can often be spent on items that operating money cannot cover. The development office will be appreciative of the help and you might gain additional benefits by keeping in touch with graduates of your programs.

Encourage more grant writing. Getting started may be hard, but once you are successful in securing grant money, your department gets better at it and learns to appreciate it. There are chairs in non-research departments who discourage grant proposals because giving faculty release time, even if paid, creates work for the chair in having to find instructors for the classes the full-time faculty would have taught. This is short-sighted thinking.

The advantages to grant money far outweigh the disadvantages. Grants can do more for your budget and hence for your department than just getting the projects or research done. You may be able to negotiate a better proportion of the salary savings or indirect costs that come back to your department; this may be especially true if the desire of the administration is for more of this activity in the department. Funds for faculty released time are usually computed as a percentage of the faculty member's salary corresponding to the proportion of released time. The replacement costs to cover the classes of the released faculty member are far below the faculty member's salary. The difference is a cost savings to the institution. Catherine Murphy reports that at her institution, 80% of the savings goes to the department and 20% to the school. Your institution might allow 100% of the savings to stay in the department budget. The added money in your personnel budget can be used on a one-time basis, and you might be able to use it for other than personnel expenditures.

Most institutions share the indirect costs recovered for grant work with the department and the college or school that houses the department. Tina Straley reports that at her former institution, 50% of the indirects went back to the department for one-time use. As department chair, she was able to use the money for faculty release time and equipment purchases. Catherine writes that she used money indirectly derived from grants for such things as a cushion for short-falls in general funds, bulk purchasing of Math Horizons for students, subscriptions to Math Intelligencer for the faculty, and travel support for students. Being grant competitive is an important argument to the dean in order to fund scholarly releases for the faculty. Catherine states that in the eyes of the faculty protecting such release time is her most important budgetary activity.

Lastly, consider laboratory fees for those classes with a laboratory component.

Stretching the Budget through Savings

The following are ways in which one might "stretch" the budget you have:

- Reduce paper expenses. Eliminate the use of department paper for students to take exams and reduce the amount used for handouts. This also reduces the photocopying bill for the department.
- Encourage the use of the Internet to reduce telephone long-distance charges.
- Travel
  - Encourage staying at less expensive hotels for meetings and conferences.
  - Encourage early registration for professional meetings. One of your faculty may get lucky and have their name drawn for a free room or other perk that

would help reduce their expenses for attending the meeting.
- Consider funding only a portion of the trip.
- Really work on finding the lowest fares on airlines. Airline fares are generally lower if reservations are made early.
- Have faculty submit proposals to external funding agencies to support their travel. In particular, if there are funding opportunities on campus, encourage faculty to submit proposals for support.

The following reflect more of a long-term effort to stretch or increase your budget through savings.

- Negotiate with your dean/provost on keeping salary savings. If you have a full professor retiring or resigning and the department and its programs will not suffer greatly, replace the full-professor with an assistant or associate professor. The savings that may come to the department will allow you to better support the entire operation.
- Increase class sizes slightly, if you can keep the salary savings. The salary savings realized may be used to provide more graders for the faculty who teach these courses, more money for travel, equipment, or whatever is needed. But be careful with this strategy; it can be a slippery slope.
- Consider some amount of part-time versus tenure-track appointments. This can be a real danger and extreme care must be given to this avenue. In fact, overuse of part-time has become a real problem in the mathematical sciences. So have a limit that you will not exceed and make that limit known to the administration.
- Make a case with the dean for support regarding "on campus" consulting. This issue can really be a factor with statisticians—the amount of statistical help expected from a committee member can in fact be quite extensive.

Making Your Case

Make good use of the MAA's Guidelines for Programs and Departments in Undergraduate Mathematical Sciences (www.maa.org/guidelines).

For comparison purposes, maintain a peer group as well as an aspirant group of departments for your institution/department. Discuss these and settle on the list in conversation with your dean and provost. Thus, when you use comparisons with these institutions in an argument, the validity of your argument has already been established.

Bibliography

1. American Association of Higher Education Website: aahebulletin.com/public/archive/archive.asp#leadership.
2. American Association of University Professors, "Contingent Faculty Appointments," *Issues in Higher Education, AAUP website*: www.aaup.org/Issues/part-time/.
3. Barone, Carol, "Conditions for Transformation: Infrastructure is Not the Issue," *EDUCAUSE Review*, May/June 2001, www.educause.edu.
4. ———, "Technology and the Changing Teaching and Learning Landscape," *AAHE Bulletin*, May, 2003, www.aahebulletin.com.
5. Boyer, Ernest L., *Scholarship Reconsidered: Priorities of the Professoriate*, San Francisco: Jossey-Bass Inc., 1990
6. Benjamin, Ernst, "How Over-Reliance on Contingent Appointments Diminishes Faculty Involvement in Student Learning," *Peer Review*, Fall 2002, Volume 5, Number 1, www.aacu.org/peerreview/pr-fa02/pr-fa02feature1.cfm.
7. Chronicle of Higher Education, *Almanac 2002–3*, chronicle.com/free/almanac/2002/index.htm.
8. Committee on the Undergraduate Program in Mathematics of the Mathematical Association of America, *Undergraduate Programs and Courses in the Mathematical Sciences: A CUPM Curriculum Guide*, Washington, DC: The Mathematical Association of America, 2004.
9. Conference Board of the Mathematical Sciences, *The Mathematical Education of Teachers, Issues in Mathematics Education, Vol. 11*, Providence, RI, American Mathematical Society in cooperation with the Mathematical Association of America, 2001. The full report (Parts 1 and 2) is available on www.cbmsweb.org. Part 1 available in print at no cost from CBMS: kolbe@math.georgetown.edu, 202-293-1170.
10. Conway, John, *On Being a Department Head, a Personal View*, Providence, RI: The American Mathematical Society, 1996.
11. DeLong, Matt, and Dale Winter, *Learning to Teach and Teaching to Learn Mathematics*, Washington, DC: The Mathematical Association of America, 2001.
12. EDUCAUSE, "2002 Campus Computing Survey," *The EDUCAUSE Guide to Evaluating Information Technology on Campus*: www.educause.edu
13. Ewing, John, editor, *Towards Excellence: Leading a Doctoral Mathematics Department in the 21st Century*, Providence, RI: The American Mathematical Society, 1999.
14. Ganter, Susan and William Barker, Editors, *Curriculum Foundation Project: Voices of the Partner Disciplines*, Washington, DC: The Mathematical Association of America, 2004.

15. Glassick, Charles E., Mary Taylor Huber, and Gene Maeroff, *Scholarship Assessed: Evaluation of the Professoriate*, San Francisco: Jossey-Bass Inc., 1997.
16. Gold, Bonnie, Sandra Keith, and William Marion, Editors, *Assessment Practices in Undergraduate Mathematics*, Washington, DC: The Mathematical Association of America, 1999.
17. Hecht, Irene W.D., Mary Lou Higgerson, Walter H. Gmelch, and Alan Tucker, *The Department Chair as Academic Leader*, Phoenix, AZ: The Oryx Press, 1999.
18. Higgerson, Mary Lou, and Susan S. Rehwaldt, *Complexities of Higher Education Administration: Case Studies & Issues,* Bolton, MA: Anker Publishing Company, Inc., 1993.
19. Leaming, Deryl R., *Academic Leadership: A Practical Guide to Chairing the Department*, Bolton, MA: Anker Publishing Company, Inc., 1998.
20. ———, *Managing People: A Guide for Department Chairs and Deans,* Bolton, MA: Anker Publishing Company, Inc., 2003.
21. Rishel, Thomas, *Teaching First: A Guide for New Mathematicians*, Washington, DC: The Mathematical Association of America, 2000.
22. Tucker, Alan, *Chairing the Academic Department: Leadership among Peers.* Phoenix, AZ: The Oryx Press, 1993.

Appendices

Resources for Department Chairs

Introduction from
CUPM Curriculum Guide 2004

Guidelines for Programs and Departments in
Undergraduate Mathematical Sciences

Resources for Department Chairs

MAA Online

We think that MAA Online (www.maa.org) offers the best place to start your search for mathematics-related information on the web

Keeping in Touch

The Combined Membership List (www.ams.org/cml)
> Supported by AMS (www.ams.org), MAA (www.maa.org), SIAM (www.siam.org), AMATYC (www.amatyc.org), AWM (www.awm-math.org) and the CMS (camel.math.ca). You can look up members using a variety of filters. Be sure to keep your own contact information current as well.

MAA Liaisons (www.maa.org/projects/liaisons/frontpage.html)
> Serve as a resource for information from the national office to their department colleagues and to respond with comments and suggestions for ways the MAA can better serve their department. MAA Sections (www.maa.org/Sections/Sections_index.html) provide a regional network to support faculty at all stages of their careers through sessions at meetings and programs such as Section NExT. See the MAA professional development pages (www.maa.org/programs) for details on these and other MAA programs.

Directory of Institutions in the Mathematical Sciences (www.ams.org/dirinst)
> Contains a list of institutions, arranged by state, which includes contact information and names of department heads.

Assessment and Your Department

The Statistical Abstract of Undergraduate Programs in the Mathematical Sciences in the United States (www.ams.org/cbms)
> Updated every five years. Available from the Conference Board for Mathematical Sciences website (www.cbmsweb.org).

Annual Survey of the Mathematical Sciences (www.ams.org/employment/deptprof.html)
> Sponsored by the AMS, MAA, ASA and Institute of Mathematical Statistics (www.imstat.org).

Towards Excellence (www.ams.org/towardsexcellence)
> Free download from the AMS.

The MAA Guidelines for Programs and Departments in the Mathematical Sciences (www.maa.org/guidelines/guidelines.html) and CUPM Guidelines (www.maa.org/cupm)
> Present recommendations that deal with a broad range of curricular and structural issues that face mathematical sciences departments and their institutional administrations.

Supporting Assessment in Undergraduate Mathematics (SAUM)
: An MAA project that offers guidelines and examples to help develop assessment programs for particular courses (or blocks of courses) and entire programs (www.maa.org/saum).

Targeted Information

CUPM Guidelines (www.maa.org/cupm)
: Offer a range of goals and examples related to specific groups of students. Background material is available through the CRAFTY Curriculum Foundations Project (www.maa.org/cupm/crafty).

The Mathematical Education of Teachers (www.cbmsweb.org/MET_Document/index.htm)
: The MET report was published by CBMS in 2001. The full report is available through the CBMS website. The MET report serves as the basis for the MAA Preparing Mathematicians to Educate Teachers (PMET) project (www.maa.org/pmet).

BIO2010: Transforming Undergraduate Education for Future Research Biologists (www.nap.edu/catalog/10497.html)
: Released by the National Research Council (www.nationalacademies.org/nrc) of the National Academies of Science (www.nas.edu) in 2002. The report is available from the National Academies Press website (www.nap.edu), along with many other reports. The MAA project Meeting the Challenges: Education Across the Biological, Mathematical and Computing Sciences (www.maa.org/mtc) provides resources for those wishing to address the significant need to reexamine quantitative training in the life sciences.

Supporting Faculty Development

The MAA PRofessional Enhancement Program (PREP)
: Serves as the umbrella for faculty development opportunities, including the PREP workshop program (www.maa.org/prep), Preparing Mathematicians to Educate Teachers (PMET) and Supporting Assessment in Undergraduate Mathematics (SAUM).

Project NExT (archives.math.utk.edu/projnext)
: Supports young faculty through both national and section level programs.

Special Interest Groups within the MAA (SIGMAAs) (www.maa.org/SIGMAA/SIGMAA.html)
: Offer a way for members with shared interests to connect with each other through special activities at regional and national meetings, and through targeted communications coordinated through the MAA.

Information on these programs as well as targeted resources to support grant writing and other professional activities are available through www.maa.org/programs as well as the Mathematics Digital Library, www.MathDL.org.

Student Resources

Information for mathematics students is available through professional society websites, including AMS, MAA and SIAM. The American Statistical Association (www.amstat.org) has information on statistical careers. The Society of Actuaries (www.soa.org) and the Casualty Actuarial Society (www.casact.org) sponsor the Be An Actuary website (www.beanactuary.org).

The Project for Nonacademic Employment (www.ams.org/careers)
: Sponsored by the AMS, MAA and SIAM, offers a variety of career information .

Career profiles (www.maa.org/students/career.html)
: More career information, including information on obtaining the MAA Career Profiles brochure, *We Do Math!* (www.maa.org/careers/brochure.html), is available on the MAA Student Career and Employment Resources page.

Mathematical Competitions
: Information on the **Putnam Exam**, for undergraduates (www.maa.org/awards/putnam.html), and **American Mathematics Competitions**, for middle and high school students, (www.unl.edu/amc) is available through MAA Online. The Consortium for Mathematics and its Applications (www.comap.com) sponsors both the **Mathematical Contest in Modeling** and the **Interdisciplinary Contest in Modeling**.

The MAA Student Chapter program (www.maa.org/students/chapter_index.html)
: Supports local faculty effort to involve students in mathematics activities outside the classroom. Student paper sessions are sponsored by the MAA Committee on Undergraduate Activities and Chapters at MathFest, with travel grants available to Student Chapter members. The undergraduate student poster sessions at the Joint Math Meetings offer another opportunity for students to participate in national meetings. The MAA Undergraduate Mathematics Conferences (www.maa.org/UGConf) program provides support for regional conferences that provide significant opportunities for students to present their work to their peers.

Undergraduate Programs and Courses in the Mathematical Sciences: CUPM Curriculum Guide 2004

A summary of the report by the
Committee on the Undergraduate Program in Mathematics of
The Mathematical Association of America

CUPM Writing Team
William Barker
David Bressoud
Susanna Epp
Susan Ganter
Bill Haver
Harriet Pollatsek (chair)

Published and Distributed by
The Mathematical Association of America

Committee on the Undergraduate Program in Mathematics (CUPM)

    Dora Ahmadi, Morehead State University
    Thomas Banchoff, Brown University
*William Barker, Bowdoin College
*Lynne Bauer, Carleton College
*Thomas Berger (Chair 1994–2000), Colby College
    David Bressoud, Macalester College
*Amy Cohen, Rutgers University
    Lynda Danielson, Albertson College
*Susanna Epp, DePaul University
*Naomi Fisher, University of Illinois at Chicago
*Joseph Gallian, University of Minnesota-Duluth
    Ramesh Gangolli, University of Washington
*Frank Giordano, US Military Academy (ret) and COMAP
    Jose Giraldo, Texas A&M University, Corpus Christi
*Bill Haver, Virginia Commonwealth University
    Dianne Hermann, University of Chicago
*Peter Hinman, University of Michigan
*Herbert Kasube, Bradley University
    Daniel Maki, Indiana University
    Joseph Malkevitch, York College CUNY
    Mercedes McGowen, Harper Community College
    Harriet Pollatsek (Chair 2000–2003), Mount Holyoke College
    Marilyn Repsher, Jacksonville University
    Allan Rossman, California Polytechnic State University, San Luis Obispo
    Kathleen Snook, US Military Academy (ret)
*Olaf Stackelberg, Kent State University
    Michael Starbird, University of Texas

* Term on CUPM ended before completion of *CUPM Guide2004*

Curriculum Project Steering Committee

William Barker
Thomas Berger
David Bressoud
Susanna Epp
Susan Ganter
Bill Haver
Herbert Kasube
Harriet Pollatsek (chair)

    Susan Ganter chairs the CUPM subcommittee on Curriculum Renewal Across the First Two Years (CRAFTY); William Barker is former chair of CRAFTY.

    The writing team acknowledges with gratitude the assistance of Barry Cipra (preliminary drafting), Kathleen Snook (compiling and editing Illustrative Resources during the winter and spring of 2003), and Tom Rishel and Michael Pearson (providing MAA staff support).

Contents

Executive Summary 143

Introduction 145

Part I: Recommendations for Departments, Programs, and all
Courses in the Mathematical Sciences 149

Part II: Additional Recommendations Concerning Specific Student Audiences 153

    A. Students taking general education or introductory collegiate courses
in the mathematical sciences 153

    B. Students majoring in partner disciplines 155

    C. Students majoring in the mathematical sciences 156

    D. Mathematical sciences majors with specific career goals 157

Executive Summary

The Mathematical Association of America's Committee on the Undergraduate Program in Mathematics (CUPM) is charged with making recommendations to guide mathematics departments in designing curricula for their undergraduate students. CUPM began issuing reports in 1953, updating them at roughly 10-year intervals. *Undergraduate Programs and Courses in the Mathematical Sciences: CUPM Curriculum Guide 2004* is based on four years of work,[1] including extensive consultation with mathematicians and members of partner disciplines.[2] Available at www.maa.org/cupm/, *CUPM Guide 2004* contains the recommendations unanimously approved by CUPM in January 2003.

Many recommendations in *CUPM Guide 2004* echo those in previous CUPM reports, but some are new. In particular, previous reports focused on the undergraduate program for mathematics majors, although with a steadily broadening definition of the major. *CUPM Guide 2004* addresses the *entire* college-level mathematics curriculum, for *all* students, even those who take just one course. *CUPM Guide 2004* is based on six fundamental recommendations for departments, programs and all courses in the mathematical sciences. The MAA Board of Governors approved these six recommendations at their Mathfest 2003 meeting.

Recommendation 1: *Mathematical sciences departments should*

- *Understand the strengths, weaknesses, career plans, fields of study, and aspirations of the students enrolled in mathematics courses;*
- *Determine the extent to which the goals of courses and programs offered are aligned with the needs of students as well as the extent to which these goals are achieved;*
- *Continually strengthen courses and programs to better align with student needs, and assess the effectiveness of such efforts.*

Recommendation 2: *Every course should incorporate activities that will help all students progress in developing analytical, critical reasoning, problem-solving, and communication skills and acquiring mathematical habits of mind. More specifically, these activities should be designed to advance and measure students' progress in learning to*

- *State problems carefully, modify problems when necessary to make them tractable, articulate assumptions, appreciate the value of precise definition, reason logically to conclusions, and interpret results intelligently;*
- *Approach problem solving with a willingness to try multiple approaches, persist in the face of difficulties, assess the correctness of solutions, explore examples, pose questions, and devise and test conjectures;*

---

[1] Supported by the National Science Foundation and the Calculus Consortium for Higher Education.

[2] Reports from a series of workshops on the mathematics curriculum with members of partner disciplines are contained in *The Curriculum Foundations Project: Voices of the Partner Disciplines,* edited and with an introduction by Susan Ganter and William Barker (MAA, 2004).

- *Read mathematics with understanding and communicate mathematical ideas with clarity and coherence through writing and speaking.*

Recommendation 3: *Every course should strive to*
- *Present key ideas and concepts from a variety of perspectives;*
- *Employ a broad range of examples and applications to motivate and illustrate the material;*
- *Promote awareness of connections to other subjects (both in and out of the mathematical sciences) and strengthen each student's ability to apply the course material to these subjects;*
- *Introduce contemporary topics from the mathematical sciences and their applications, and enhance student perceptions of the vitality and importance of mathematics in the modern world.*

Recommendation 4: *Mathematical sciences departments should encourage and support faculty collaboration with colleagues from other departments to modify and develop mathematics courses, create joint or cooperative majors, devise undergraduate research projects, and possibly team-teach courses or units within courses.*

Recommendation 5: *At every level of the curriculum, some courses should incorporate activities that will help all students progress in learning to use technology*
- *Appropriately and effectively as a tool for solving problems;*
- *As an aid to understanding mathematical ideas.*

Recommendation 6: *Mathematical sciences departments and institutional administrators should encourage, support and reward faculty efforts to improve the efficacy of teaching and strengthen curricula.*

Part I of *CUPM Guide 2004* elaborates on these recommendations and suggests ways that a department can evaluate its progress in meeting them. Part II contains supplementary recommendations concerning particular student audiences:

A. Students taking general education or introductory courses in the mathematical sciences;
B. Students majoring in partner disciplines, including those preparing to teach mathematics in elementary or middle school;
C. Students majoring in the mathematical sciences;
D. Mathematical sciences majors with specific career goals: secondary school teaching, entering the non-academic workforce, and preparing for post-baccalaureate study in the mathematical sciences and allied disciplines.

Specific methods for implementation are not prescribed, but the online document *Illustrative Resources for CUPM Guide 2004* at www.maa.org/cupm/ describes a variety of experiences and resources associated with these recommendations. These illustrative examples are not endorsed by CUPM, but they may serve as a starting point for departments considering enhancement of their programs. Pointers to additional resources, such as websites (with active links) and publications, are also given.

Introduction

Mathematics is universal: it underlies modern technology, informs public policy, plays an essential role in many disciplines, and enchants the mind. At the start of the twenty-first century, the undergraduate study of mathematics can and should be a vital and engaging part of preparation for many careers and for well-informed citizenship. In the *CUPM Guide 2004*, the term 'mathematics' is generally synonymous with 'mathematical sciences' and refers to a collection of mathematics-related disciplines, including, but not necessarily limited to, pure and applied mathematics, mathematics education, computational mathematics, operations research, and statistics. Departments of mathematical sciences can and should play a central role in their institutions' undergraduate programs. The *CUPM Guide 2004*[3] calls on mathematicians and mathematics departments to rethink the full range of their undergraduate curriculum and co-curriculum to ensure the best possible mathematical education for all their students, from liberal arts students taking just one course to students majoring in the mathematical sciences.

The need for action

Over the past one hundred years mathematics has become more important to more disciplines than ever before. At the same time both the number and diversity of students in post-secondary education and the variety of their mathematical backgrounds have increased dramatically. Additionally, computer technology has forever altered the way mathematics is used in the workplace, from retail store registers to financial institutions to laboratories doing advanced scientific research.

These developments present unprecedented curricular challenges to departments of mathematical sciences—challenges many departments and individual faculty members are engaged in meeting. During the past twenty years there has been an explosive increase in the number of presentations and publications on issues and innovations in the teaching of post-secondary mathematics.[4] This activity reflects a growing movement to address the undergraduate mathematics curriculum conscientiously and creatively.

But there are indicators that all is not well. National data provide clear evidence that undergraduate mathematics programs are under serious pressure, with decreasing numbers of mathematics majors and declining enrollment in advanced mathematics courses.[5] From 1985 to 2000 the total number of bachelor's degrees awarded annually in the U.S. rose 25% and the number of science and technology degrees grew

---

[3] The MAA publication *Guidelines for Programs and Departments in Undergraduate Mathematical Sciences*, MAA, 2001, available at www.maa.org/guidelines/guidelines.html, complements the *CUPM Guide 2004* and other curricular reports by presenting a set of recommendations that deal with a broad range of structural issues that face mathematical sciences departments and their institutional administrations.

[4] For example, at the January 2003 Joint Mathematics Meetings, more than one third of the talks concerned mathematics education.

[5] Appendix 3 of the full *Guide* contains further analysis of data on numbers of majors and the supply of secondary teachers of mathematics, and Appendix 4 contains data on enrollment in and availability of advanced courses.

20%. However, data collected by the Conference Board of the Mathematical Sciences (CBMS)[6] show that the total number of degrees awarded annually by mathematics and statistics departments, including those in secondary mathematics education, stayed essentially flat during this 15-year period. In fact, the annual total in these departments fell 4% between 1995 and 2000, and the number of annual degrees in mathematics fell 19% in the 1990s. The drop in mathematics degrees occurred at the same time as the need for new teachers of secondary mathematics grew more acute.

One might expect the increase in science and technology degrees to translate into higher enrollment in advanced mathematics courses as allied subjects. In fact, the opposite has occurred: enrollment in advanced courses taught in mathematics departments has fallen, dropping 25% from 1985 to 2000; an increase from 1995 to 2000, while encouraging, has not returned enrollment to 1990 levels. Further, CBMS data show that even the *availability* of advanced courses has declined in the past five years, as the percentages of departments offering several typical courses[7] has decreased, in some cases by more than 20%. This trend is unfortunate, not only for the health of the mathematical sciences major but also because the health of disciplines that use mathematics—and by extension the health of society—is enhanced when a significant number of students are knowledgeable about the advanced mathematics that is relevant to their fields.

Striking successes at a number of colleges and universities demonstrate that these trends can be reversed. For instance, the MAA volume *Models That Work: Case Studies in Effective Undergraduate Mathematics Programs*[8], summarizes effective practices at a set of mathematics departments that have excelled in (i) attracting and training large numbers of mathematics majors, or (ii) preparing students to pursue advanced study in mathematics, or (iii) preparing future school mathematics teachers, or (iv) attracting and training underrepresented groups in mathematics. Site-visits to ten departments and information on a number of others revealed "no single key to a successful undergraduate program in mathematics." However, there were common features. "What was a bit unexpected was the common attitude in effective programs that the faculty are not satisfied with the current program. They are constantly trying innovations and looking for improvement."

Areas for attention and action

Mathematics departments need to serve *all* students well—not only those who major in the mathematical or physical sciences. The following steps will help departments reach this goal.

- Design undergraduate programs to address the broad array of problems in the diverse disciplines that are making increasing use of mathematics.
- Guide students to learn mathematics in a way that helps them to better understand its place in society: its meaning, its history, and its uses. Such understanding is often lacking even among students who major in mathematics.
- Employ a broad range of instructional techniques, and require students to confront, explore, and communicate important ideas of modern mathematics and the uses of mathematics in society. Students need more classroom experiences in which they learn to think, to do, to analyze—not just to memorize and reproduce theories or algorithms.

---

[6] *CBMS 2000: Statistical Abstract of Undergraduate Programs in the Mathematical Sciences in the United States*, D. Lutzer, J. Maxwell and S. Rodi, AMS, 2002.

[7] Including algebra, analysis, geometry, mathematical modeling and applied mathematics; see Table 4-3 in Appendix 4 of the full *Guide*.

[8] MAA Notes **38** (1995), Alan C. Tucker, editor.

- Understand and respond to the impact of computer technology on course content and instructional techniques.
- Encourage and support faculty in this work—a task both for departments and for administrations.

The *CUPM Guide 2004* presents six general recommendations to assist mathematics departments in the design and teaching of all of their courses and programs. Later in the *Guide* these recommendations are elaborated and made specific for particular student audiences.

Using the *CUPM Guide 2004*

Part I of the *CUPM Guide 2004* elaborates on and specifies the meaning of the six general recommendations as well as suggesting ways that a department can evaluate progress in meeting them. Part II contains supplementary recommendations for particular student audiences.

Some students major in fields that do not require specific mathematical preparation. They may take one course in mathematics, perhaps to satisfy a general education requirement of their institution or major program. Section A of Part II addresses the needs of these students, many of whom—especially among the hundreds of thousands enrolled each semester in courses called College Algebra—are not optimally served by the mathematics courses they take.

Partner disciplines are those whose majors are required to take one or more specific mathematics courses. These disciplines vary by institution but usually include the physical sciences, the life sciences, computer science, engineering, economics, business, education, and often several social sciences. Recommendations concerning these students are in Section B of Part II, including those for pre-service K–8 teachers.

Section C of Part II contains recommendations concerning students majoring in the mathematical sciences. The recommendations urge departments to learn the probable career paths and needs of their majors and offer them a flexible program that provides appropriate breadth and depth. Section D contains further recommendations for mathematical sciences majors preparing to teach secondary school mathematics, planning for non-academic employment, or intending post-baccalaureate study.

There are many ways to carry out each recommendation, and different choices will be appropriate in different institutional settings. Consequently, these recommendations rarely specify particular courses or syllabi. That doesn't mean "anything goes." Indeed, each recommendation is accompanied by measures to help a department gauge its effectiveness.[9] As stated in the *MAA Guidelines for Programs*, "These measures will, of necessity, be multi-dimensional since no single statistic can adequately represent departmental performance with respect to most departmental goals. Measures of student learning and other student outcomes should be included."[10] Course syllabi and sample assignments, along with their contribution to students' grades, are other valuable measures.

Although no specific methods for implementation are outlined, the section entitled Illustrative Resources is designed to help departments implement and improve practices to satisfy the recommendations. It is organized and numbered the same way as the recommendations in Parts I and II. A variety of examples, including assignments, courses (with suggested syllabi and texts), and programs are provided for each recommendation. The examples range along a continuum, from modest first steps and small changes that can be easily effected to more ambitious efforts. Pointers to additional resources, such as websites and publications, are also given.

These recommendations have been reduced to a core judged essential for building and supporting department strength and effectively meeting department obligations. They are not a wish list for an ideal

---

[9] Appendix 6 of the full *Guide* contains sample questions for department self-study.

[10] MAA, 2001, available at www.maa.org/guidelines/guidelines.html.

future department. Indeed, the reality is that departments at many institutions are coping with diminished human and financial resources and conflicting and escalating demands on faculty time. Moreover, meaningful change is never easy. Nonetheless, the use of the word "should" in a recommendation means that departments are expected to make a conscientious effort to achieve steady improvement until they are able to satisfy it.

Background for the recommendations

The Mathematical Association of America's Committee on the Undergraduate Program in Mathematics (CUPM) is charged with making recommendations to guide mathematics departments in designing curricula for their undergraduate students. CUPM began issuing reports in 1953, updating them at roughly 10-year intervals. In 1999 work began on the current recommendations. CUPM solicited position papers from prominent mathematicians and conducted panel discussions and focus groups at national meetings to obtain reactions to preliminary drafts of these recommendations. There has been extensive consultation with other professional societies in the mathematical sciences. From 1999 to 2002 CUPM's subcommittee on Curriculum Renewal Across the First Two Years (CRAFTY) conducted a series of workshops on the mathematics curriculum with participants from a broad range of partner disciplines.[11]

The six fundamental recommendation of this *Guide* were approved by the Board of Governors of the MAA in July 2003. All of the recommendations were unanimously approved by CUPM in January 2003. Many of the current recommendations echo those in previous CUPM reports, but some are new. In particular, previous reports focused on the undergraduate program for mathematics majors, although with a steadily broadening definition of the major in the 1981 and 1991 reports. The *CUPM Guide 2004*, in contrast, addresses the *entire* college-level mathematics curriculum, for *all* students, even those who take just one course.[12]

---

[11] Appendices 1 and 2 of the full *Guide* contain detailed accounts of CUPM's activities and of the CRAFTY workshops collectively known as the Curriculum Foundations project. The results of the project are contained in the MAA publication *The Curriculum Foundations Project: Voices of the Partner Disciplines,* edited and with an introduction "A Collective Vision: Voices of the Partner Disciplines" by Susan Ganter and William Barker.

[12] While attempting to address the college-level curriculum in mathematics more comprehensively, the *CUPM Guide 2004* does not discuss a number of important issues, including non-credit or developmental courses and articulation between institutions.

Part I: Recommendations for Departments, Programs, and all Courses in the Mathematical Sciences

1. Understand the student population and evaluate courses and programs

In summarizing the common features of the programs described in *Models That Work,* the authors wrote that one of the "states of mind that underlie faculty attitudes in effective programs" is "teaching for the students one has, not the students one wished one had." *Towards Excellence: Leading a Doctoral Mathematics Department in the 21st Century* echoes this theme: "Mathematics departments should position themselves to receive new or reallocated resources by meeting the needs of their institutions. That does not mean sacrificing the intellectual integrity of an academic program, nor does it mean relegating mathematics to a mere service role. It *does* mean fulfilling a bargain with the institution in which one lives, and for most departments a major part of that bargain involves instruction."[13]

Recommendation 1: Mathematical sciences departments should
- Understand the strengths, weaknesses, career plans, fields of study, and aspirations of the students enrolled in mathematics courses;
- Determine the extent to which the goals of courses and programs offered are aligned with the needs of students as well as the extent to which these goals are achieved;
- Continually strengthen courses and programs to better align with student needs, and assess the effectiveness of such efforts.

2. Develop mathematical thinking and communication skills

The power of mathematical thinking — pattern recognition, generalization, abstraction, problem solving, careful analysis, rigorous argument — is important for every citizen. It is highly valued by employers and by other disciplines but widely misunderstood and undervalued by students.

Communication is integral to learning and using mathematics, and skill in communicating is commonly listed as the most important quality employers seek in a prospective employee.[14] However, many students expect mathematics classes to be wordless islands where they won't be asked to read, write, or discuss ideas.

Appropriate instructional approaches to reasoning and proof have been passionately debated among mathematicians for decades, but with a greater sense of urgency during the last twenty years. While much remains to be learned about how best to teach reasoning and proof skills—as well as how best to improve

---

[13] American Mathematical Society Task Force on Excellence, J. Ewing editor, AMS, 1999, p. xiii.

[14] See, for instance, surveys by the National Association of Colleges and Employers, www.naceweb.org.

communication skills—a variety of strategies can help students progress. Mathematics faculty should deliver an unambiguous message concerning the importance of mathematical reasoning and communication skills and adopt instructional methods and curriculum content that develop these skills. Designing a curriculum that develops these skills effectively and at appropriate levels for all students is one of the biggest and most important challenges for mathematics departments.

Recommendation 2: Every course should incorporate activities that will help all students progress in developing analytical, critical reasoning, problem-solving, and communication skills and acquiring mathematical habits of mind. More specifically, these activities should be designed to advance and measure students' progress in learning to

- State problems carefully, modify problems when necessary to make them tractable, articulate assumptions, appreciate the value of precise definition, reason logically to conclusions, and interpret results intelligently;
- Approach problem solving with a willingness to try multiple approaches, persist in the face of difficulties, assess the correctness of solutions, explore examples, pose questions, and devise and test conjectures;
- Read mathematics with understanding and communicate mathematical ideas with clarity and coherence through writing and speaking.

3. Communicate the breadth and interconnections of the mathematical sciences

Many students do not see the connections between mathematics and other disciplines or between mathematics and the world in which they live. Too often they leave mathematics courses with a superficial mastery of skills that they are unable to apply in non-routine settings and whose importance to their future careers is unrecognized. Conceptual understanding of mathematical ideas and facility in mathematical thinking are essential for both applications and further study of mathematics, yet they are often lost in a long list of required topics and computational techniques. Even when students successfully apply mathematical techniques to problems, they are often unable to interpret their results effectively or communicate them with clarity.

The beauty, creativity, and intellectual power of mathematics and its contemporary challenges and discoveries, are often unknown and unappreciated. The interplay between differing perspectives—continuous and discrete, deterministic and stochastic, algebraic and geometric, exact and approximate—is appreciated by very few students, even though flexible use of these varying perspectives is critical for applications and for learning new mathematics.

Recommendation 3: Every course should strive to
- Present key ideas and concepts from a variety of perspectives;
- Employ a broad range of examples and applications to motivate and illustrate the material;
- Promote awareness of connections to other subjects (both in and out of the mathematical sciences) and strengthen each student's ability to apply the course material to these subjects;
- Introduce contemporary topics from the mathematical sciences and their applications, and enhance student perceptions of the vitality and importance of mathematics in the modern world.

4. Promote interdisciplinary cooperation

Mathematics programs have traditionally drawn heavily from the physical sciences for applications. In recent years, mathematics has come to play a significant role in far more disciplines, but many mathematics programs have not adjusted to this new reality.

Mathematics departments should seize the opportunity to harness the growing awareness in other disciplines of the power and importance of mathematical methods. A curriculum developed in consultation with other disciplines that includes a variety of courses and degree options can attract more students, help them learn important mathematical ideas, retain more students for intermediate and advanced coursework, strengthen their ability to apply mathematics to other areas, and improve the quantity and quality of the mathematics majors and minors.

Recommendation 4: Mathematical sciences departments should encourage and support faculty collaboration with colleagues from other departments to modify and develop mathematics courses, create joint or cooperative majors, devise undergraduate research projects, and possibly team-teach courses or units within courses.

5. Use computer technology to support problem solving and to promote understanding

Recent advances in desktop and handheld computer technology can be used to improve the pedagogy and content of mathematics courses at all levels. Some mathematical ideas and procedures have become less important because of these emerging technological tools; others have gained importance. The 2001 *MAA Guidelines for Programs and Departments in Undergraduate Mathematical Sciences* recommended that departments "should employ technology in ways that foster teaching and learning, increase the students' understanding of mathematical concepts, and prepare students for the use of technology in their careers or in their graduate study."[15] However, few mathematics departments have effectively met the challenges posed by the growth of technology, and many are only beginning to address seriously the issues it raises.

Recommendation 5: At every level of the curriculum, some courses should incorporate activities that will help all students progress in learning to use technology
- Appropriately and effectively as a tool for solving problems;
- As an aid to understanding mathematical ideas.

6. Provide faculty support for curricular and instructional improvement

Many of the recommendations in this *Guide*, including collaborating with colleagues in other disciplines, adapting material from other parts of mathematics or from other disciplines for use in teaching, evaluating student writing, and making effective use of technology, require time and effort from faculty beyond what they might ordinarily devote to the revision and creation of courses. Departments and administrators need to acknowledge that meeting these recommendations makes substantial demands on faculty (and, in some cases, on graduate teaching assistants and other temporary or part-time instructors).

Recommendation 6: Mathematical sciences departments and institutional administrators should encourage, support and reward faculty efforts to improve the efficacy of teaching and strengthen curricula.

---

[15] *Guidelines for Programs and Departments in Undergraduate Mathematical Sciences*, MAA, 2001, available at www.maa.org/guidelines/guidelines.html.

Part II: Additional Recommendations Concerning Specific Student Audiences

The recommendations in Part II of the *Guide* are supplementary to those in Part I addressing all students.

A. Students taking general education or introductory collegiate courses in the mathematical sciences

General education and introductory courses enroll almost twice as many students as all other mathematics courses combined.[16] They are especially challenging to teach because they serve students with varying preparation and abilities who often come to the courses with a history of negative experiences with mathematics. Perhaps most critical is the fact that these courses affect life-long perceptions of and attitudes toward mathematics for many students—and hence many future workers and citizens. For all these reasons these courses should be viewed as an important part of the instructional program in the mathematical sciences.

This section concerns the student audience for these entry-level courses that carry college credit. A large percentage of these students are enrolled in college algebra. Traditional college algebra courses, with a primary emphasis on developing skills in algebraic computation, have a long history at many institutions.

Students enrolled in college algebra courses generally fall into three categories:

1. Students taking mathematics to satisfy a requirement but not specifically required to take a course called college algebra;
2. Students majoring in areas or studying within states or university systems that specifically require a course called college algebra;
3. Students intending to take courses such as statistics, calculus, discrete mathematics, or mathematics for prospective elementary or middle school teachers and who need additional preparation for these courses.

Unfortunately, there is often a serious mismatch between the original rationale for a college algebra requirement and the actual needs of the students who take the course. A critically important task for mathematical sciences departments at institutions with college algebra requirements is to clarify the rationale for the requirements, determine the needs of the students who take college algebra, and ensure that the department's courses are aligned with these findings (see Recommendation A.2).

---

[16] According to the CBMS study in the Fall of 2000, a total of 1,979,000 students were enrolled in courses it classified as "remedial" or "introductory" with course titles such as elementary algebra, college algebra, pre-calculus, algebra and trigonometry, finite mathematics, contemporary mathematics, quantitative reasoning. The number of students enrolled in these courses is much greater than the 676,000 enrolled in calculus I, II or III, the 264,000 enrolled in elementary statistics, or the 287,000 enrolled in all other undergraduate courses in mathematics or statistics. At some institutions, calculus courses satisfy general education requirements. Although calculus courses can and should meet the goals of Recommendation A.1, such courses are not the focus of this section.

Because many students taking introductory mathematics decide not to continue to higher level courses, general education and introductory courses often serve as students' last exposure to college mathematics. It is important, therefore, that these courses be designed to serve the future mathematical needs of such students as well as to provide a basis for further study for students who do continue in mathematics. All students, those for whom the course is terminal and those for whom it serves as a springboard, need to learn to think effectively, quantitatively and logically. Carefully-conceived courses—described variously as quantitative literacy, liberal arts mathematics, finite mathematics, college algebra with modeling, and introductory statistics—have the potential to provide all the students who take them with the mathematical experiences called for in this section.

A common feature of many effective courses and programs that have been developed for these students is the leadership provided by key faculty members. It requires committed and talented faculty to understand the needs of these students and the opportunities inherent in these courses. Continuing leadership is needed and special training must be provided for instructors—including graduate assistants and part-time faculty—to offer courses that will meet the needs of these students.

A.1: Offer suitable courses

All students meeting general education or introductory requirements in the mathematical sciences should be enrolled in courses designed to

- Engage students in a meaningful and positive intellectual experience;
- Increase quantitative and logical reasoning abilities needed for informed citizenship and in the workplace;
- Strengthen quantitative and mathematical abilities that will be useful to students in other disciplines;
- Improve every student's ability to communicate quantitative ideas orally and in writing;
- Encourage students to take at least one additional course in the mathematical sciences.

A.2: Examine the effectiveness of college algebra

Mathematical sciences departments at institutions with a college algebra requirement should

- Clarify the rationale for the requirement and consult with colleagues in disciplines requiring college algebra to determine whether this course—as currently taught— meets the needs of their students;
- Determine the aspirations and subsequent course registration patterns of students who take college algebra;
- Ensure that the course the department offers to satisfy this requirement is aligned with these findings and meets the criteria described in A.1.

A.3: Ensure the effectiveness of introductory courses

General education and introductory courses in the mathematical sciences should be designed to provide appropriate preparation for students taking subsequent courses, such as calculus, statistics, discrete mathematics, or mathematics for elementary school teachers. In particular, departments should

- Determine whether students that enroll in subsequent mathematics courses succeed in those courses and, if success rates are low, revise introductory courses to articulate more effectively with subsequent courses;
- Use advising, placement tests, or changes in general education requirements to encourage students to choose a course appropriate to their academic and career goals.

B. Students majoring in partner disciplines

Partner disciplines vary by institution but usually include the physical sciences, the life sciences, computer science, engineering, economics, business, education, and often several social sciences.[17] It is especially important that departments offer appropriate programs of study for students preparing to teach elementary and middle school mathematics. Recommendation B.4 is specifically for these prospective teachers.

B.1 Promote interdisciplinary collaboration

Mathematical sciences departments should establish ongoing collaborations with disciplines that require their majors to take one or more courses in the mathematical sciences. These collaborations should be used to

- Ensure that mathematical sciences faculty cooperate actively with faculty in partner disciplines to strengthen courses that primarily serve the needs of those disciplines;
- Determine which computational techniques should be included in courses for students in partner disciplines;
- Develop new courses to support student understanding of recent developments in partner disciplines;
- Determine appropriate uses of technology in courses for students in partner disciplines;
- Develop applications for mathematics classes and undergraduate research projects to help students transfer to their own disciplines the skills learned in mathematics courses;
- Explore the creation of joint and interdisciplinary majors.

B.2: Develop mathematical thinking and communication

Courses that primarily serve students in partner disciplines should incorporate activities designed to advance students' progress in
- Creating, solving, and interpreting basic mathematical models;
- Making sound arguments based on mathematical reasoning and/or careful analysis of data;
- Effectively communicating the substance and meaning of mathematical problems and solutions.

B.3: Critically examine course prerequisites

Mathematical topics and courses should be offered with as few prerequisites as feasible so that they are accessible to students majoring in other disciplines or who have not yet chosen majors. This may require modifying existing courses or creating new ones. In particular,
- Some courses in statistics and discrete mathematics should be offered without a calculus prerequisite;
- Three-dimensional topics should be included in first-year courses;
- Prerequisites other than calculus should be considered for intermediate and advanced non-calculus-based mathematics courses.

B.4: Pre-service elementary (K–4) and middle school (5–8) teachers

Mathematical sciences departments should create programs of study for pre-service elementary and middle school teachers that help students develop

---

[17] Appendix 2 of the full *CUPM Guide 2004* contains a list of the disciplines represented at the Curriculum Foundations workshops.

- A solid knowledge—at a level above the highest grade certified—of the following mathematical topics: number and operations, algebra and functions, geometry and measurement, data analysis and statistics and probability;
- Mathematical thinking and communication skills, including knowledge of a broad range of explanations and examples, good logical and quantitative reasoning skills, and facility in separating and reconnecting the component parts of concepts and methods;
- An understanding of and extensive experience with the uses of mathematics in a variety of areas;
- The knowledge, confidence, and motivation to pursue career-long professional mathematical growth.

C. Students majoring in the mathematical sciences

The recommendations in this section refer to all major programs in the mathematical sciences, including programs in mathematics, applied mathematics, and various tracks within the mathematical sciences such as operations research or statistics. Also included are programs designed for prospective mathematics teachers, whether they are "mathematics" or "mathematics education" programs, although requirements in education are not specified in this section.

Although these recommendations do not specifically address minors in the mathematical sciences, departments should be alert to opportunities to meet student needs by creating minor programs—for example, for students preparing to teach mathematics in the middle grades.

These recommendations also provide a basis for discussion with colleagues in other departments about possible joint majors with any of the physical, life, social or applied sciences.

C.1: Develop mathematical thinking and communication skills

Courses designed for mathematical sciences majors should ensure that students

- Progress from a procedural/computational understanding of mathematics to a broad understanding encompassing logical reasoning, generalization, abstraction and formal proof;
- Gain experience in careful analysis of data;
- Become skilled at conveying their mathematical knowledge in a variety of settings, both orally and in writing.

C.2: Develop skill with a variety of technological tools

All majors should have experiences with a variety of technological tools, such as computer algebra systems, visualization software, statistical packages, and computer programming languages.

C.3: Provide a broad view of the mathematical sciences

All majors should have significant experience working with ideas representing the breadth of the mathematical sciences. In particular, students should see a number of contrasting but complementary points of view:

- Continuous and discrete,
- Algebraic and geometric,
- Deterministic and stochastic,
- Theoretical and applied.

Majors should understand that mathematics is an engaging field, rich in beauty, with powerful applications to other subjects, and contemporary open questions.

C.4: Require study in depth

All majors should be required to
- Study a single area in depth, drawing on ideas and tools from previous coursework and making connections, by completing two related courses or a year-long sequence at the upper level;
- Work on a senior-level project that requires them to analyze and create mathematical arguments and leads to a written and an oral report.

C.5: Create interdisciplinary majors

Mathematicians should collaborate with colleagues in other disciplines to create tracks within the major or joint majors that cross disciplinary lines.

C.6: Encourage and nurture mathematical sciences majors

In order to recruit and retain majors and minors, mathematical sciences departments should
- Put a high priority on effective and engaging teaching in introductory courses;
- Seek out prospective majors and encourage them to consider majoring in the mathematical sciences;
- Inform students about the careers open to mathematical sciences majors;
- Set up mentoring programs for current and potential majors, and offer training and support for any undergraduates working as tutors or graders;
- Assign every major a faculty advisor and ensure that advisors take an active role in meeting regularly with their advisees;
- Create a welcoming atmosphere and offer a co-curricular program of activities to encourage and support student interest in mathematics, including providing an informal space for majors to gather.

D. Mathematical sciences majors with specific career goals

D.1: Majors preparing to be secondary school (9–12) teachers

In addition to acquiring the skills developed in programs for K–8 teachers, mathematical sciences majors preparing to teach secondary mathematics should
- Learn to make appropriate connections between the advanced mathematics they are learning and the secondary mathematics they will be teaching. They should be helped to reach this understanding in courses throughout the curriculum and through a senior-level experience that makes these connections explicit;
- Fulfill the requirements for a mathematics major by including topics from abstract algebra and number theory, analysis (advanced calculus or real analysis), discrete mathematics, geometry, and statistics and probability with an emphasis on data analysis;
- Learn about the history of mathematics and its applications, including recent work;
- Experience many forms of mathematical modeling and a variety of technological tools, including graphing calculators and geometry software.

D.2: Majors preparing for the nonacademic workforce

In addition to the general recommendations for majors, programs for students preparing to enter the nonacademic workforce should include
- A programming course, at least one data-oriented statistics course past the introductory level, and coursework in an appropriate cognate area; and
- A project involving contemporary applications of mathematics or an internship in a related work area.

D.3: Majors preparing for post-baccalaureate study in the mathematical sciences and allied disciplines

Mathematical sciences departments should ensure that
- A core set of faculty members are familiar with the master's, doctoral and professional programs open to mathematical sciences majors, the employment opportunities to which they can lead, and the realities of preparing for them;
- Majors intending to pursue doctoral work in the mathematical sciences are aware of the advanced mathematics courses and the degree of mastery of this mathematics that will be required for admission to universities to which they might apply. Departments that cannot provide this coursework or prepare their students for this degree of mastery should direct students to programs that can supplement their own offerings.

The Mathematical Association of America

Guidelines
for Programs and Departments in Undergraduate Mathematical Sciences

Revised Edition

February 2003

Contents

A. Introduction 163

B. Planning and Periodic Review 164

C. Program Faculty and Staffing 165
    1. Educational Background 165
    2. Promoting Excellence in Teaching 165
    3. Promoting Excellence in Scholarship 166
    4. Promoting Excellence in Service 167
    5. Assignment of Duties 167
    6. Adequate Staffing Levels 168
    7. Securing and Sustaining a Diverse Faculty 168
    8. Faculty Evaluation and Rewards 168
    9. Support Staff 169

D. Curriculum and Teaching 169
    1. Curriculum Planning and Review Procedures (See Section B, items 1, 2, 3) ... 169
    2. Curriculum Access and Pedagogy 170
    3. Quantitative Reasoning for College Graduates 170
    4. Program Recommendations of Professional Societies 170
    5. Research in Teaching and Learning 171
    6. Impact of Technology 171

E. Resources 172
    1. Faculty Resources 172
    2. Office Space 172
    3. Classroom Equipment 172
    4. Computer Resources 172
    5. Informal Gathering Space for Majors 172
    6. Library Facilities 172

F. Students 172
    1. Advising 172
    2. Broadening the Student Base in the Mathematical Sciences 173
    3. Co-curricular Activities 173

References 173

Appendix A
    Quantitative Reasoning for College Graduates: A Complement to the Standards ... 177

Appendix B
   The Undergraduate Major in the Mathematical Sciences 179
Appendix C
   A Call for Change: Recommendations for the
      Mathematical Preparation of Teachers of Mathematics 180
Appendix D
   Providing Resources for Computing in Undergraduate Mathematics 182
Appendix E
   Advising 183
Appendix F
   Task Force to Review the 1993 MAA Guidelines for
      Programs and Departments in Undergraduate Mathematical Sciences 184

A. Introduction

In 1989, the National Research Council published the report, "Everybody Counts: A Report to the Nation on the Future of Mathematics Education," [37]. That report characterized undergraduate mathematics as the "linchpin for revitalization of mathematics education" and reminded us that "critical curricular review and revitalization take time, energy, and commitment."

Since 1989, several other reports of national organizations included recommendations for strengthening this linchpin role of undergraduate mathematics. These documents include "Moving Beyond Myths: Revitalizing Undergraduate Mathematics," [38]; "Challenges for College Mathematics: An Agenda for the Next Decade," [6]; "The Undergraduate Major in the Mathematical Sciences," [17]; "A Call for Change: Recommendations for the Mathematical Preparation of Teachers of Mathematics," [28]; "Principles and Standards for School Mathematics," [35]; "Heeding the Call for Change: Suggestions for Curricular Action," [52]; and "Recognition and Rewards in the Mathematical Sciences," [26]. The Guidelines that follow incorporate many of these recommendations.

This document supports the many curricular reports by presenting a set of recommendations that deal with a broad range of structural issues that face mathematical sciences departments and their institutional administrations. The document includes statements on planning and periodic review, faculty and staffing, curriculum and teaching, institutional and departmental resources, physical facilities, libraries, and services to students such as advising and cocurricular activities for majors. As such, these Guidelines deal with all aspects of the undergraduate mission—general education, mathematical sciences courses serving other disciplines, and the mathematical sciences education of majors, including future secondary teachers. They are intended to address mathematical sciences programs in four-year colleges and universities but many of the guidelines also apply to two-year colleges. In June 1993, the American Mathematical Association of Two-Year Colleges (AMATYC) published the document, "Guidelines for Mathematics Departments at Two-Year Colleges," [57], which is a complement to this document. The AMATYC guidelines extend these Guidelines to two-year colleges, and they address more specifically the concerns of these institutions. The AMATYC Guidelines may be found at the AMATYC web address www.amatyc.org/publications.html/.

These Guidelines are intended to be used by mathematical sciences programs in self-studies, planning, and assessment of their undergraduate programs, as well as by college and university administrators and external reviewers. Mathematical sciences programs and their administrations can use the recommendations included in these Guidelines as a basis for allocating resources and planning for the future. It is the joint responsibility of institutional administrations and mathematical sciences programs to provide and use properly the resources necessary to meet these Guidelines.

In many institutions, faculty from more than one mathematical sciences discipline are in a department or program that also includes pure and applied mathematics. These Guidelines apply to all programs in such a mathematical sciences department. Such programs might include pure mathematics, applied mathematics, mathematics education, computer science, statistics, and operations research. In some institutions, academic programs may not be organized into traditional department units. These Guidelines are intended to apply to the mathematical sciences courses, programs, and faculty within those institutions as well. Application of these Guidelines to programs in separate departments of computer science, operations research, statistics, or mathematics education is not intended.

Throughout the document, the phrase "mathematical sciences" refers to a collection of mathematics-related disciplines, including, but not necessarily limited to, pure and applied mathematics, mathematics education, computer science and computational mathematics, operations research, and statistics. In this document, we use the word "department(s)" to include non-departmental mathematical sciences programs.

We urge that this Guidelines document be used as the starting point for the planning and evaluation process. Professional societies in the mathematical sciences can provide advice for the review process and on the selection of external reviewers. An excellent reference for a planning and evaluation process is the text, "Towards Excellence: Leading a Mathematics Department in the 21st Century," [22]. Though written for mathematics departments with doctoral programs, most of its chapters provide valuable information, conclusions, and advice for the review and evaluation process in all collegiate mathematical sciences departments. Chapter 20, in particular, "How to Conduct External Reviews," applies directly to the process. The text focuses on precisely the kinds of academic concerns and issues confronting mathematical sciences departments, their department and college leadership, and the faculty who deliver the programs.

Another source of information is the publication of the Association of American Colleges, "Program Review and Educational Quality in the Major," [7]. Important comparative data can be found in the MAA publication, "Statistical Abstract of Undergraduate Programs in the Mathematical Sciences in the United States, Fall 1995 CBMS Survey," [31].

Comparative data also can be found in the report of the AMSASA-IMS-MAA Data Committee's Annual Survey of the Mathematical Sciences, published in three parts each year in the "Notices of the American Mathematical Society." The Fall 2000 CBMS Survey Report is published on the AMS website, e-MATH, and the MAA website, MAA Online. The Data Committee's Annual Survey reports also are published on e-MATH. For departments with graduate programs, the annual AMS publication, "Assistantships and Graduate Fellowships in the Mathematical Sciences," [4] gives valuable comparative information, not published elsewhere, about departments. This publication also is available at the e-MATH website.

Other resource documents that relate directly to the departmental planning and evaluation process include, "Reinventing Undergraduate Education: A Blueprint for America's Research Universities," [12], "The SIAM Report on Mathematics in Industry," [34], "Factors Contributing to High Attrition Rates Among Science, Mathematics and Engineering Majors," [46], "Twenty Questions that Deans Should Ask Their Mathematics Departments," [49], and "Models That Work: Case Studies in Effective Undergraduate Programs," [61]. As cited in the References Section, the first two of these reports are available on the internet.

B. Planning and Periodic Review

1. In cooperation with the dean or other appropriate administrative official, each mathematical sciences department should participate at regular intervals in a process of periodic planning and evaluation. Participants in the process should include faculty, students, alumni, client departments, external mathematical sciences reviewers, and deans or other administrators. The faculty and any external consultants directly involved in this review should adequately reflect both the program mission and the faculty of the mathematical sciences program or department being reviewed. The process should lead to a strategic plan, acceptable to the department and to its dean, for enhancing strengths and remedying deficiencies identified in the planning and evaluation process.

2. The major components of the planning and evaluation process should be:
   a. A statement that clearly defines the mission of the undergraduate mathematical sciences department.
   b. A delineation of the educational goals of the program as well as a statement of how attainment of these goals is expected to fulfill the mission of the program.
   c. Procedures for measuring the extent to which the educational goals are being met. These measures will, of necessity, be multi-dimensional since no single statistic can adequately represent departmental performance with respect to most departmental goals. Measures of student learning and other student outcomes should be included in the procedures.
   d. A process for regularly reviewing (and revising, if necessary) departmental and academic program components in light of measurements of program success.
   e. A departmental and institutional plan to allocate, over time, the resources needed to implement the strategic plan agreed to by the department and its dean.

3. The periodic reviews should examine all aspects of the department's undergraduate academic program. Reviewers should consider the departmental mission and goals statements, faculty and staffing issues, the extent to which the department's curriculum is consistent with those statements and with the needs of the students being served, evidence that indicates the extent to which the department's service courses give students the mathematical sciences background they need to take subsequent courses in other departments, evidence that indicates the extent to which the department's major program is successful in enabling students to meet the department's educational goals, the effectiveness of the department's advising practices, and the success of the department in recruiting and retaining students, including students from groups that are underrepresented in the current personnel pool of mathematical scientists. Curricular quality and effectiveness should be judged in comparison with mathematical sciences programs at peer departments, and in comparison with the most recent CUPM recommendations on the mathematics curriculum.

When related to the department's goals, further indicators of program quality may include student performance in seminars, departmental comprehensive examinations, course-embedded assessment, undergraduate research activities, internships, consulting experiences, and national competitions and examinations (such as the COMAP modeling competition, the Putnam competition, the Data Analysis competition, actuarial examinations, and the GRE mathematics examination). Other indicators include student evaluations that are obtained through surveys and interviews. Reviewers should also consider the accomplishments of the graduates of the department's programs and, where appropriate, the number of mathematical sciences majors produced compared to peer departments and to national averages, the success of the associate degree recipients who transfer to four-year colleges, the success of bachelor's degree recipients who matriculate in post-graduate degree programs, and the employability of the department's associate or bachelor's graduates. Reviewers should also address institutional and departmental resources, including physical facilities and library resources.

C. Program Faculty and Staffing

1. Educational Background

a. Except as indicated in item c below, those who are hired to teach mathematical sciences courses for undergraduate credit should have a minimum of a master's degree in a mathematical science. This applies to both full-time and part-time faculty wherever the institution's courses are taught. In institutions that grant at least a bachelor's degree in the mathematical sciences, tenure-track faculty should possess a doctoral degree in a mathematical science.

b. Mathematical sciences departments frequently offer courses in several disciplines, including pure mathematics, mathematics education, applied mathematics, computer science, operations research, and statistics. Ideally, a course should be taught by a faculty member with a graduate degree in the discipline of the course. In the many departments where this is not possible, the course should have a developer/coordinator, who has a graduate degree in that discipline. The developer/coordinator should hold regular meetings with the faculty teaching the course in order to discuss such items as the course syllabus, textbooks, resources, teaching methods, technical matters, and evaluation. The department's curricular needs should be a major factor in departmental hiring decisions. The number of faculty with expertise in a mathematical sciences discipline should reflect the department's courses and enrollments in that discipline. See also Guideline C.2.e.v and Guideline D.1.g.

c. Since they are the future faculty members of our colleges and universities, it is important that graduate students have some instruction in teaching including serving as apprentice teachers. Thus, even though they might not meet the requirements above, mathematical sciences graduate students may teach or assist with the teaching of courses under the close supervision of faculty members. A graduate student who has a master's degree or equivalent in a mathematical science may be assigned as the independent instructor of record in a course. As is the case with all faculty teaching in the department, unless the graduate student's master's degree or equivalent is in the same discipline as the course, the course coordinator should consult regularly with the graduate student. In addition, the graduate student should be provided with the same resources for teaching that are available to full-time faculty teaching the same course, including office space, computer and library resources, and mentoring by full-time faculty. Other activities that are suitable for graduate teaching assistants include grading papers, staffing laboratories, conducting discussion or recitation sections, and tutoring.

d. If undergraduate students assist in undergraduate instruction, their efforts should be restricted to classroom organizational duties such as collecting papers; reading and commenting on homework assignments; tutoring or assisting in mathematics and computer laboratories, mathematics workshops, and recitation sections; and holding supplementary instruction sessions.

2. Promoting Excellence in Teaching

a. Teaching ability and commitment to teaching should be key factors in all appointments to the teaching staff.

b. Orientation and training programs should be provided to familiarize new staff members with departmental expectations and the needs of students. New faculty should receive a description of the teaching and teaching-related duties expected of them and the means by which those duties will be evaluated.

c. Faculty should be supervised, monitored, and evaluated in order to help them improve their teaching. See also Guideline C.8.f.

d. The courses assigned to faculty, especially those newly hired, should be chosen to aid in their development as teachers.
e. A regular program for maintaining and improving teaching expertise is essential for all academic mathematical scientists.
   i. Departments should provide long-term structured opportunities for acquisition and improvement of teaching skills by all who teach. This might be accomplished through demonstrations of pedagogical approaches and strategies for good teaching and may include videotaping and peer critiques, observing classes taught by outstanding teachers, team teaching with these teachers, or working with faculty mentors.
   ii. Departments should provide regular opportunities for and support the professional development of faculty members to learn of the most recent findings about teaching and learning in the mathematical sciences and of the most recent developments in technology that support teaching and learning. See also Guideline D.6.
   iii. When a department decides to use technology in a course or program, it should offer appropriate training for faculty in that technology and its effective use in instruction.
   iv. All full-time faculty members should participate regularly in activities to maintain and improve their teaching expertise. A suggested outline for improvement can be found in the CTUM report, "A Source Book for College Mathematics Teaching," [44] and in the first nine pages of "A Call for Change: Recommendations for the Mathematical Preparation of Teachers of Mathematics," [28].
   v. Participation in programs designed to assist college teachers is particularly important for members of a department who sometimes teach outside of their own mathematical sciences discipline. These programs should be extensive in scope and require substantial investment of time by participants. Many faculty have found that earning a master's degree or equivalent in a second mathematical sciences discipline or other discipline appropriate to the teaching assignment gives them the needed background.
   vi. When instituting programs for the improvement of teaching by graduate teaching assistants and part-time instructors, consideration should be given to the characteristics of the model programs and the remarks presented in the introduction of the MAA publication, "Keys to Improved Instruction by Teaching Assistants and Part-time Instructors," [14].
f. In certain circumstances, part-time faculty can make unique contributions to a mathematical sciences department. Departments that employ part-time instructors should provide them with all of the resources necessary for teaching that are provided to full-time instructors, including office space as well as computer, Internet, and library resources. Full-time faculty should mentor part-time faculty in resolving problems, in meeting responsibilities, and in familiarizing them with the procedures and expectations of the department. See also the last two sentences of Guideline C.6.
g. Departments should ensure that senior faculty assume a leadership role in the undergraduate program by participating fully in teaching, curriculum development, and student advising. In addition, they have key responsibility for reviewing and nurturing junior faculty and teaching assistants.
h. Both senior and junior faculty should, at least on occasion, teach courses at all levels of the undergraduate program.

3. Promoting Excellence in Scholarship

a. All full-time faculty members should, as part of their work assignments, engage in disciplinary or interdisciplinary scholarship, broadly defined to include the discovery of new knowledge, the integration of knowledge, the application of knowledge, and scholarship related to teaching, "Scholarship Reconsidered: Priorities of the Professoriate," [11]. Successful scholarship includes the obligation of timely communication of results to peers. Faculty should sustain their scholarship throughout their careers. Guidelines for the acceptable forms of this scholarship and for the nature of communication of results to peers should be made available in writing to faculty members. A department should encourage, recognize, and value the diverse nature of faculty scholarship that is directly related to the department's mission and program goals.
b. A regular program for maintaining and improving disciplinary or interdisciplinary expertise is essential for all academic mathematical scientists. Departments should support professional development of faculty members to enable them to remain current with the most recent advances in the field. Appropriate devel-

opment opportunities include participation in seminars, graduate level mathematical sciences courses, appropriate courses in other disciplines, conferences, symposia, short courses, and professional meetings. As all full-time faculty members should participate in appropriate professional development, such activities should be a part of each faculty member's work assignment. Sabbaticals, other faculty leave programs, faculty exchanges, and periodic workload reductions provide faculty the necessary time for professional development.

c. Mentoring programs and faculty development opportunities designed specifically for new faculty should be available, and all new faculty should be encouraged to participate in such activities. See Guideline C.8.f.

d. In order to foster a sustained commitment to scholarship among faculty, departments and their institutions should provide sabbatical or research leaves at appropriate intervals and should have generous policies allowing leaves without pay for research and scholarly activities.

4. Promoting Excellence in Service

a. Departments should expect senior faculty to seek and accept committee assignments within the department, the institution, and the profession. Departments should expect junior faculty to become involved in service, at a level consistent with local expectations for tenure, with the understanding that faculty governance responsibilities increase upon the award of tenure. See also Guidelines C.7.b. iii and iv.

b. Departments should expect all full-time faculty members to be formally involved in their professions by participating in professional organizations.

5. Assignment of Duties

In this section, "hours" will mean semester hours, and a "course" will be considered to carry three semester credit hours. Appropriate adjustments should be made for quarter hours, for labs, or for courses carrying credit hours other than three.

a. Institutional and departmental missions vary considerably. Work assignments for faculty should reflect institutional and departmental missions. They should be consistent with locally defined expectations for promotion and tenure as well as with comparisons to assignments in peer departments at other institutions.

   i. Faculty for whom personnel decisions are based primarily upon assessment of substantial scholarly accomplishments or doctoral level teaching and research supervision should have teaching assignments that do not exceed two courses per semester.

   ii. Faculty for whom personnel decisions are based upon assessment of contributions in teaching, scholarship, and service should have teaching assignments that reflect these multiple expectations and allow for attention to non-classroom responsibilities. Teaching assignments above three courses per semester, when combined with other faculty responsibilities, do not allow the time needed to develop and maintain a program of sustained scholarship with the result that tenure and promotion might be effectively unattainable. For such faculty, teaching assignments above the level of three courses per semester must be avoided.

   iii. Faculty for whom personnel decisions are based predominantly upon assessment of teaching and service responsibilities must have sufficient time for class preparation, course development, conducting office hours, advising, and other duties in service of the profession in addition to formal classroom teaching. Teaching assignments that exceed five courses or a maximum of three different class preparations or fifteen contact hours do not allow sufficient time for these responsibilities.

b. Depending on department or program mission and priorities, appropriate reductions from the normal teaching assignments described above should be made for extensive involvement in professional activities or service. This may include such activities or service as committee or administrative assignments; course, courseware, program, or computational technology development; laboratory supervision; thesis direction; and scholarship.

c. In the assignment of duties, departments must exercise careful monitoring of an individual faculty member's total responsibility to the program. Total responsibility for a large number of students in a single course or supervision of course assistants can add as much to work assignments as an additional course. In making teaching assignments, departments must take into account not only the number of contact hours assigned, but also the number of students enrolled in those classes and, if teaching assistants are used, any additional supervisory responsibility.

d. A valuable part of the professional duties of some mathematical sciences department faculty members is the use of their expertise in providing professional consulting for their institution. The institution and the faculty member should place in writing an agreement describing exactly what the institution expects from the faculty member in these professional consulting activities that are not part of the workloads of all faculty members. The agreement should describe how the consulting activities will be evaluated, how they will be considered in the tenure and promotion process, and how they fit in the faculty member's work assignment. See also Guideline C.8.e.

e. The institution and the department should have a public written policy on the amount of time that a full time faculty member can spend during the academic year on outside activities for compensation.

6. Adequate Staffing Levels

Department staffing levels should be sufficient to allow personal interaction between student and instructor to occur in all courses, to give tenure-track faculty adequate time to meet tenure expectations, to allow faculty to engage in scholarship consistent with departmental expectations, and to meet work assignment expectations similar to those of peer departments at comparable institutions. Many mathematical sciences programs today tend to have too large a percentage of part-time faculty, and, over time, should convert part-time positions into full-time positions. This fosters the participation of a greater percentage of faculty in the work of the department.

7. Securing and Sustaining a Diverse Faculty

a. The mathematical sciences are in constant need of being strengthened and replenished by drawing well-educated individuals from the broadest possible pool of talent. It is essential to widen the spectrum from which mathematical sciences faculty are drawn. Members of traditionally underrepresented groups, including women, minorities, the physically challenged and those from educationally deprived backgrounds, deserve special attention in this regard. The first step toward widening the talent pool from which new faculty are drawn is to make certain that all new positions are advertised in places seen by all potential faculty members.

b. Hiring decisions are only first steps in achieving and sustaining a diverse faculty. Subsequent issues of faculty development are equally important.
   i. A department should maintain an atmosphere that welcomes all people who seek to work and study in the mathematical sciences disciplines in that department.
   ii. Departments have a special responsibility to newly hired faculty from historically underrepresented groups (see above) to protect them from excessive demands on their time and energy from advising and committee service that go beyond what is expected of other faculty members.
   iii. Departments recruiting faculty from historically underrepresented groups must accept the responsibility for nurturing the professional growth and advancement of these faculty, especially during their early years of employment, in order to insure long-term diversity rather than short-term.
   iv. Departments should be on record as endorsing and enforcing the institution-approved personnel policies, including policies on non-discrimination and sexual and other harassment.

8. Faculty Evaluation and Rewards

a. The department should have written procedures for evaluating its faculty members on the basis of teaching, scholarship, and service. These departmental procedures should be made available to all departmental faculty and should be reviewed periodically.

b. Tenure-track, non-tenured faculty should be counseled annually as to progress toward tenure.

c. "Departments should use the best available methods, imperfect though they may be, for evaluating teaching, scholarship, and service while also seeking to develop better methods of evaluation." See "Guiding Principle V" on page 35 in "Recognition and Rewards in the Mathematical Sciences," [26].

d. "Every institution and department should work to develop efficient, robust, reliable, and trusted measures of teaching effectiveness. These could include peer evaluation, surveying of students from current and previous semesters (graduating seniors or alumni, for example), studying student achievement in subsequent courses, reviewing syllabi and examinations, and other techniques." See "Discussion" to "Guiding Principle V" on page 35 in "Recognition and Rewards in the Mathematical Sciences" [26]. Also see the Mathematical Sciences Education Board document, "Report of the Task Force on Teaching Growth and Effectiveness," [39].

e. In accordance with departmental mission and priorities, some consulting and other professional activities may advance the scholarship and teaching of faculty members and the department. Consulting and other professional activities may fit into the category of teaching or scholarship and in that case should be evaluated accordingly, or such activities might be evaluated as a separate category, with correspondly less emphasis on other categories. Supervision processes and evaluation procedures for formal consulting activities should include the monitoring of faculty progress in maintaining and improving the quality of these activities. Evaluation criteria and procedures for consulting activities must be a part of a written agreement among the faculty member, the department, and the appropriate dean.

f. Professional expectations vary considerably among the mathematical sciences disciplines. When a department has faculty members from several disciplines, it is particularly important that there be a mutually accepted, written statement concerning expectations for the faculty members in the areas of teaching, scholarship, and service, and, if relevant, consulting. It is important that the agreed upon expectations statement be the basis for personnel decisions. Departments should consult position papers of various professional societies in preparing such expectations statements. Furthermore, if the department has only one or two faculty members in a discipline, it should seek outside persons to serve as advisors for departments and mentors for these isolated faculty members early in their careers. If such outside advisors or mentors are used, it is important that they and the department give the same messages to the faculty member about departmental and institutional expectations. Professional societies can identify senior faculty members who are willing to serve as outside advisors and mentors.

g. "Each department must develop a rewards system consonant with its own mission and the mission of the institution. In formulating a rewards structure, each department must analyze who its constituencies are, what they need from the department, and whether those needs are being met." See "Discussion" to "Guiding Principle VI," p. 37, in "Recognition and Rewards in the Mathematical Sciences," [26].

9. Support Staff

Clerical and technical staff should be sufficient to support the teaching and scholarly activities of the department. It is particularly important to have adequate technical staff to maintain computers used by students, faculty, and clerical staff. Faculty should not be expected to provide computer support for the department.

D. Curriculum and Teaching

1. Curriculum Planning and Review Procedures (see Section B, items 1, 2, 3)

a. The department should have the primary responsibility and most influential voice in setting the placement policies, the prerequisites or co-requisites, the course content, and the exit competencies for the department's courses. See Guideline F.1.a.

b. Departments should discuss with client departments plans to change mathematical sciences courses or programs in ways that would have significant effects on academic programs in the client departments. This consultation should continue throughout the process of making the change. See Guideline F.1.a.

c. There should be established procedures for periodic review of the curriculum. These reviews, which should be a part of the duties of faculty assigned by the department, should include careful scrutiny of course syllabi, prerequisites, and textbooks. These reviews should examine the curriculum in the context of the departmental goals and institutional mission. They should include consideration of the curriculum's relevance and appropriateness for the students being served. Effective reviews often lead to revision, addition, or deletion of courses.

d. Many courses within mathematical sciences programs are organized with a sequence of prerequisites. Course prerequisites should be clearly stated and equitably enforced. A current syllabus for each course should be on the web and on file for review by faculty colleagues and by students. Catalog course descriptions should be kept up-to-date. Departments should take the necessary steps to ensure that all sections of a given course are consistent in content, use of technology, focus, and rigor.

e. In cases where the department regularly teaches students who transfer from two-year colleges, the department should cooperate with those two-year colleges in facilitating student transfers. Mathematical sciences faculty members at the institutions should work together to ensure compatibility of appropriate courses, and course equivalencies should be published. Faculty should ensure that the courses taught at the two- and four-year colleges are consistent in content, technology, focus, and rigor.

f. The development and review process for courses that support other programs should involve faculty members from those programs. In addition, informal contacts with faculty from other departments can provide useful information concerning the mathematical sciences courses that their students must take. Working collaborations with faculty colleagues in departments of education must be established to strengthen the programs that prepare teachers of school mathematics.

g. In cases where a department offers a course or courses in a particular discipline, but does not have a faculty member with expertise in that discipline, the department should take special care to consult the curricular guidelines of the relevant professional society in that discipline.

2. Curriculum Access and Pedagogy

a. The mathematical sciences curriculum should be responsive to the needs of the department's students. Course and program offerings should provide suitable academic challenge and should be based on the expectation that all students can learn mathematics. The spectrum of beginning courses should be broad enough to offer appropriate choices and placement in mathematics for all students entering the institution.

b. Departments must be provided with the resources necessary to deliver high quality teaching that includes the opportunity for students to interact frequently and nontrivially with their instructors. Departments should facilitate these personal interactions by avoiding the use of large lecture settings that require students to become passive audiences. The best way to encourage active student faculty interactions and to enable faculty to give students individual attention is to provide a small-class environment with fewer than thirty students in each section. Also with restricted class size, faculty members gain flexibility to adopt a teaching style that best fits both the material to be learned and their students' needs.

c. The instructional staff assigned to each course should be sufficient to allow for regular and frequent feedback to students about their progress. Feedback from instructors should take various forms, such as critical reviews of short quizzes and hour tests, comments and suggestions for homework or writing assignments, and critiques of students' presentations of projects and contributions in seminars. Interaction in classes, mathematics laboratories, and workshops provides additional feedback. Instructors should consider all of these forms of evaluation not just as evaluation of the students but also as information that can be used to improve their teaching. Instructors can gain information for improving their teaching also from student journals and mid-semester questionnaires.

d. Courses which are required in a student's program of study but have a history of low enrollment should be scheduled and taught at least once every two years regardless of the low enrollment. Courses that are not scheduled at least every two years should not be listed in the college catalog.

3. Quantitative Reasoning for College Graduates

In 1996, the MAA Board of Governors approved the report, "Quantitative Reasoning for College Graduates: A Complement to the Standards," [15]. The summary and preface of this report may be found in Appendix A of this document. The text of the full report may be found at the MAA Online Web address www.maa.org/past/ql/ql_toc.html.

a. Mathematical Sciences departments should assume the responsibility of actively developing and promoting within their institutions quantitative literacy general education requirements for all undergraduates.

b. These quantitative literacy requirements

  i. Should be consistent with the Report, "Quantitative Reasoning for College Graduates: A Complement to the Standards,"

  ii. Should emphasize teaching students to use mathematical methods to solve real-world problems, and

  iii. Should involve courses at both the lower and upper division levels. The report, "Quantitative Literacy: Why Numeracy Matters for Schools and Colleges," [48] helps to provide a rationale for the role of mathematics in quantitative literacy programs.

4. Program Recommendations of Professional Societies

a. The mathematical sciences bachelor's degree program should be consistent with the current recommendations of the MAA Committee on the Undergraduate Program in Mathematics (CUPM) Guidelines. Departments should provide for majors the experiences described in the section, "Completing the Major," of the CUPM Report [17]. Programs with no curricular track that conforms to the CUPM guidelines should be justified by a detailed and persuasive rationale for departing from those guidelines. A summary of the CUPM Report comprises Appendix B of this document. The full report has been reprinted in "Heeding the Call for Change: Suggestions for Curricular Action" [52], pp. 225–247. Those who develop or deliver statistics major, minor, or concentration programs within mathematical sciences programs or departments should know the recommendations contained in the American Statistical Association report, "Curriculum Guidelines for Undergraduate Programs in Statistical Science," [3].

b. If the institution offers a program of study leading to certification of elementary or secondary mathematics teachers, that program should be consistent with the current guidelines of the MAA Committee on the Mathematical Education of Teachers (COMET), "A Call for a Change: Recommendations for the Mathematical Preparation of Teachers of Mathematics," [28]. In addition, the faculty who develop and deliver the program should know the recommendations contained in the Conference Board on Mathematical Sciences report, " The Mathematical Education of Teachers," [18] (the CBMS MET Report) and in the National Council of Teacher of Mathematics (NCTM) publication, "Principles and Standards for School Mathematics,"[35]. A summary of the COMET Report is contained in Appendix C of this document. Those who develop and deliver two-year college school teacher preparation programs should know the recommendations contained in the National Science Foundation report, "Investing in Tomorrow's Teachers: The Integral Role of Two-Year Colleges in the Science and Mathematics Preparation of Prospective Teachers," [40].

5. Research in Teaching and Learning

Departments should be aware of the results of research on teaching and learning in the mathematical sciences, and they should make use of those results in improving instruction. Such research can provide a useful framework for such pedagogical matters as what students know and can do, how they develop their understanding of mathematical concepts, how they solve problems, how various kinds of mathematics teaching affects learning, and how students read proofs. No one method of instruction is optimal for all students, for all faculty members, or for all subject matter. Departments should encourage and assist faculty members who investigate, try out, and evaluate alternative teaching techniques that show promise in helping some students be more successful in learning in the mathematical sciences. See also Guideline C.2.e.ii.

6. Impact of Technology

Mathematical sciences departments should employ technology in ways that foster teaching and learning, increase the students' understanding of mathematical concepts, and prepare students for the use of technology in their careers or their graduate study. Where appropriate, courses offered by the department should integrate current technology. The availability of new technological tools and their pervasive use in the workplace have the potential for changing both the curriculum and the way that the mathematical sciences can and should be taught. See also Guideline C.2.e.iii.

a. Departments should review and adjust the curriculum to reflect the expanded use of technology in each discipline and in the workplace.

b. In courses where the use of modern technology will enhance student learning, departments should adopt methods of teaching mathematical sciences courses that make full use of appropriate current technology. These methods include laboratory sessions and assignments using computer software or graphing calculators, electronic communication with students, demonstrations in class using projection equipment, group activities fostered by technology, and use of the Internet.

c. The student activities and experiences related to technology should be designed primarily to enhance the learning of mathematics and may serve to introduce the students to mathematically related technology. Special emphasis should be placed on giving prospective teachers the experience of learning mathematics using, or adapting, methods practiced in the schools and on educating these prospective teachers to be leaders in the effective use of technology in the schools.

d Faculty should consider the many different ways to employ technology in order to foster interaction among instructors and students.

e. Departments should develop a general policy for assessment of student work that acknowledges the role of technology in the curriculum. In particular, departments should

   i. Adopt a policy of testing students in the way that they actually do their coursework. That is, if students regularly use graphing calculators or computer software for assignments, then the same facilities should be available during tests. If desired, students may be tested separately for computational facility and particular facts without the use of technology.

   ii. Adopt a policy on assessment of technology-based student projects and assignments.

E. Resources

1. Faculty Resources

All faculty should be furnished with the resources (computers, computer software, and travel funds, for example) necessary for them to perform the teaching and the scholarly activities they were hired to do. It is critical in this regard that new faculty have access to these resources when they begin work in their faculty positions.

2. Office Space

All full-time mathematics faculty members should have private offices. Each part-time faculty member should have a desk and office space that allows confidential conferences with students outside the classroom.

3. Classroom Equipment

Classrooms should be equipped with such traditional teaching aids as adequate board space and projector equipment and screens. Other teaching aids such as computer and internet access, CDROMs, slide and video projection equipment, and computer projectors and monitors should be available as needed.

4. Computer Resources

a. The department's access to computer resources for teachers and students should be consistent with the MAA policy statement contained in Appendix D.
b. The department faculty should have high-speed access to the Internet. This access should enable faculty to quickly download graphics and data.

5. Informal Gathering Space for Majors

There should be dedicated space for use by mathematical sciences majors for conversation and study. It is desirable that this space be near faculty offices to allow opportunity for frequent contact between students and faculty.

6. Library Facilities

a. Library holdings should include the publications labeled "Essential," "Highly Recommended," or "Recommended," at the MAA Internet page, "Basic Library List" for undergraduate mathematics. The Internet address for the page is www.maa.org/data/bll/home.htm.
b. Library holdings should be sufficient to provide mathematics enrichment materials for undergraduate student projects and to meet the scholarly needs of the program faculty. If specific library materials are not available on site, then they should be readily available through a process of interlibrary loan.
c. The institution's libraries should be staffed, scheduled, and located in such a way that their mathematical sciences holdings are readily available to all faculty members and students.
d. Library holdings of books and periodicals in the mathematical sciences should be reviewed periodically by committees, which include representatives of the mathematical sciences departments.

F. Students

1. Advising

a. Departments should have established policies and procedures for placement in introductory mathematical sciences courses. It is important that these policies be well understood and disseminated across the institution.
b. The mathematical sciences faculty, admissions personnel, and other freshman/sophomore advisors should periodically review the effectiveness of these placement procedures for entering freshmen.
c. Advisement of transfer students should involve cooperative efforts of mathematical sciences faculty of both sending and receiving institutions.
d. Departments should make information available to all students about their educational programs in the mathematical sciences through printed and internet media (such as an advising handbook) and through informal and formal advising.
e. Departments should provide majors and other students with information about careers in the mathematical sciences and should make qualified students aware of further educational opportunities. In particular, qualified students should be encouraged to take advantage of summer programs at other institutions including programs that provide undergraduate research experience.
f. Every student who declares a major in the mathematical sciences should be assigned an advisor from the mathematical sciences faculty. Advisors should take an active role in meeting regularly with their advisees, particularly with students who seem reluctant to ask questions. The CUPM Statement on Advising, contained in Appendix E, provides a model for departments to follow in their advising programs.

2. Broadening the Student Base in the Mathematical Sciences

The nation's work force is becoming increasingly dependent on substantial mathematical preparation, and demographics indicate that women and minorities constitute a growing percentage of new entrants to the work force. Department faculty must address a changing student body whose experiences, cultural backgrounds, and learning styles may be significantly different from those of current mathematical sciences faculty members.

It is essential to ensure that women, underrepresented minorities, and students from educationally deprived backgrounds are encouraged both to study the mathematical sciences at all levels and to become part of the mathematical sciences community. Appropriate programs and courses in the mathematical sciences should be available to all students who are admitted to the institution and have an interest in mathematics. Further, in both curricular and co-curricular activities, there should be concentrated efforts on the part of the department's faculty directed to assuring that courses, programs, and the departmental climate are inviting and supportive to all students regardless of their gender or cultural background. To achieve these goals, it is recommended that departments:

a. Have explicit policies and related practices to attract and retain members of groups currently underrepresented in the mathematical sciences;

b. Distribute career information on mathematical sciences-based careers that actively encourages these choices by students, especially minorities and women;

c. Develop policies and articulation agreements with two-year colleges to facilitate student transfers between two-year and four-year institutions;

d. Work to increase the presence of women and minorities in academia and other mathematical sciences-based careers by encouraging qualified women and minority students to pursue graduate study in the mathematical sciences;

e. Initiate intervention projects whose participants include pre-college students in minority communities and work with predominantly minority organizations to encourage persistence of minority students in their study of the mathematical sciences;

f. Include more opportunities for student interaction as might take place in mathematical sciences laboratories, tutorial sessions, or structured small group learning sessions;

g. Ensure that all departmental facilities and activities are accessible to students who are physically disadvantaged; and

h. Enforce university or departmentally approved policies, including those addressing sexual harassment and discrimination, as they apply to relationships among students and between students and faculty. If relevant policies do not already exist, departments should seek, through appropriate governance channels, to establish the necessary policies needed to foster positive departmental atmospheres.

3. Co-curricular Activities

a. Department faculty should be involved with undergraduates in co-curricular activities designed to create an atmosphere of inclusion and cohesiveness among mathematical sciences majors and a sense of participation in the department. This atmosphere should be attractive to all majors but especially to women and those of diverse cultural backgrounds. Mathematical sciences clubs, honorary societies such as Kappa Mu Epsilon and Pi Mu Epsilon, MAA Student Chapters and student chapters of other professional societies are possible options. Teams for the Putnam and Modeling competitions and scheduled departmental social activities that include undergraduates are other such activities.

b. Special colloquia appropriate for undergraduates should be regularly held.

c. Departments should encourage faculty to work with undergraduate students in research projects.

References

1. Albers, Donald J.; Anderson, Richard D.; and Loftsgaarden, Don O. *Undergraduate Programs in the Mathematical and Computer Sciences: The 1985–1986 Survey*. MAA Notes 7. Washington, DC: Mathematical Association of America, 1987.

2. Albers, Donald J.; Loftsgaarden, Don O.; Rung, Donald C.; Watkins, Ann E. *Statistical Abstract of Undergraduate Programs in the Mathematical Sciences and Computer Science in the United States: The 1990–91 CBMS Survey*. MAA Notes 23. Washington, DC: Mathematical Association of America, 1992.

3. American Statistical Association. *Curriculum Guidelines for Undergraduate Programs in Statistical Science*. Alexandria, VA: American Statistical Association, 2001. Available at www.amstat.org/education/Curriculum_Guidelines.html

4. AMS-ASA-IMS-MAA Data Committee. *Assistantships and Graduate Fellowships in the Mathematical Sciences*.

Providence, RI: American Mathematical Society, 2001. Available at www.ams.org/employment/asst2001-frnt.pdf.

5. Association of American Colleges. *Integrity in the College Curriculum.* Washington, DC: Association of American Colleges, 1985.

6. Association of American Colleges Study of the Arts and Sciences Major. *Challenges for College Mathematics: An Agenda for the Next Decade.* Washington, DC: Mathematical Association of America, 1990.

7. Association of American Colleges. *Program Review and Educational Quality in the Major.* Washington, DC: Association of American Colleges, 1992.

8. Berriozabal, Manuel P. Why Hasn't Mathematics Worked for Minorities? *UME Trends,* 1:2 (May 1989) 8.

9. Board on Mathematical Sciences. *Actions for Renewing U.S. Mathematical Sciences Departments.* Washington, DC: National Research Council, 1990.

10. Board on Mathematical Sciences. *Renewing U.S. Mathematics: A Plan for the 1990s.* National Research Council. Washington, DC: National Academy Press, 1990.

11. Boyer, Ernest L. *Scholarship Reconsidered: Priorities of the Professoriate.* Princeton, NJ: The Carnegie Foundation for the Advancement of Teaching, 1990.

12. Boyer Commission for Educating Undergraduates at the Research University. *Reinventing Undergraduate Education: A Blueprint for America's Research Universities.* Stony Brook, NY: Carnegie Foundation, 1998. Available at naples.cc.sunysb.edu/Pres/boyer.nsf/.

13. Case, Bettye Anne (Ed.). How Should Mathematicians Prepare for College Teaching? *Notices of the American Mathematical Society,* 36:10 (December 1989) 1344–1346.

14. Case, Bettye Anne (Ed.). *Keys to Improved Instruction by Teaching Assistants and Part-time Instructors.* MAA Notes No. 11. Washington, DC: Mathematical Association of America, 1989.

15. Committee on the Undergraduate Program in Mathematics. Quantitative Reasoning for College Graduates: A Complement to the Standards. Report of the Subcommittee on Quantitative Literacy Requirements. Washington, DC: Mathematical Association of America, 1998.

16. Committee on the Undergraduate Program in Mathematics. *Reshaping College Mathematics.* MAA Notes 13. Washington, DC: Mathematical Association of America, 1989.

17. Committee on the Undergraduate Program in Mathematics. The Undergraduate Major in the Mathematical Sciences. Washington, D.C.: Mathematical Association of America, 1991.

18. Conference Board of Mathematical Sciences. *The Mathematical Education of Teachers.* Published for the Conference Board of Mathematical Sciences by the American Mathematical Society in Cooperation with the Mathematical Association of America. Providence, RI: American Mathematical Society, 2001. Available at www.cbmsweb.org/MET_Document/index.htm.

19. David, Edward E. Renewing U.S. Mathematics: An Agenda to Begin the Second Century. *Notices of the American Mathematical Society,* 35 (October 1988) 1119–1123.

20. Douglas, Ronald G. (Ed.). *Toward a Lean and Lively Calculus.* MAA Notes 6. Washington, DC: Mathematical Association of America, 1986.

21. Duren, W.L., Jr. *A General Curriculum in Mathematics for Colleges.* Washington, DC: Mathematical Association of America, 1965.

22. Ewing, John H. (Ed.). *Towards Excellence: Leading a Mathematics Department in the 21st Century.* American Mathematical Society Task Force on Excellence. Providence, RI: American Mathematical Society, 1999. Available at www.ams.org/towardsexcellence/.

23. Gillman, Leonard. Teaching Programs That Work. *Focus,* 10:1 (1990) 710.

24. Gopen, George D., and Smith, David A. What's an Assignment Like You Doing in a Course Like This? Writing to Learn Mathematics. Reprinted in *The College Mathematics Journal,* 21 (1990) 219.

25. Halmos, Paul R. The Calculus Turmoil. *Focus,* 10:6 (Nov.–Dec. 1990) 13.

26. Joint Policy Board for Mathematics. *Recognition and Rewards in the Mathematical Sciences, Report of the Committee on Professional Recognition and Rewards.* Providence, RI: American Mathematical Society, 1994.

27. Kenschaft, Patricia Clark. *Winning Women into Mathematics.* Washington, DC: Mathematical Association of America, 1991.

28. Leitzel, James R.C. (Ed.). *A Call for Change: Recommendations for the Mathematical Preparation of Teachers of Mathematics.* Washington, DC: Mathematical Association of America, 1991.

29. Leitzel, James, R.C.; Tucker, Alan. *Assessing Calculus Reform Efforts.* Washington, DC: Mathematical Association of America, 1995.

30. Loftsgaarden, Don O.; Rung, Donald C.; Watkins, Ann E. *Statistical Abstract of Undergraduate Programs in the Mathematical Sciences in the United States. Fall 1995 CBMS Survey.* MAA Reports Number 2. Washington, DC: Mathematical Association of America, 1997.

31. Lutzer, David J.; Maxwell, James W.; Rodi, Stephen B. *Statistical Abstract of Undergraduate Programs in the Mathematical Sciences in the United States. Fall 2000 CBMS Survey.* Providence, RI: American Mathematical Society, 2002. Available at www.ams.org/cbms/.

32. Madison, Bernard L., and Hart, Therese A. *A Challenge of Numbers: People in the Mathematical Sciences.* Washington, DC: National Academy Press, 1990.

33. Mathematical Association of America. *Mathematical Scientists at Work: Careers in the Mathematical Sciences.* Washington, DC: Mathematical Association of America, 1991.

34. Mathematics in Industry Steering Committee. *The SIAM Report on Mathematics in Industry.* Philadelphia, PA:

Society for Industrial and Applied Mathematics, 1995. Available at www.siam.org/mii/miihome.htm.

35. National Council of Teachers of Mathematics. *Principles and Standards for School Mathematics.* Reston, VA: National Council of Teachers of Mathematics, 2000.

36. National Council of Teachers of Mathematics. *Professional Standards for Teaching Mathematics.* Reston, VA: National Council of Teachers of Mathematics, 1991.

37. National Research Council. *Everybody Counts: A Report to the Nation on the Future of Mathematics Education.* Washington, DC: National Academy Press, 1989.

38. National Research Council. Moving *Beyond Myths: Revitalizing Undergraduate Mathematics.* Washington, DC: Committee on Mathematical Sciences in the Year 2000, National Academy Press, 1991.

39. National Research Council. Report of the Task Force on Teaching Growth and Effectiveness. Washington, DC: National Academy Press, 1993.

40. National Science Foundation. Investing in Tomorrow's Teachers: The Integral Role of Two-Year colleges in the Science and Mathematics Preparation of Prospective Teachers. Washington, DC: National Science Foundation, 1999. Available at www.nsf.gov/pubs/1999/nsf9949/nsf9949.txt.

41. National Science Foundation. *Women and Minorities in Science and Engineering.* Washington, DC: National Science Foundation, 1988.

42. Oaxaca, Jaime, and Reynolds, Ann W. (Eds.) *Changing America: The New Face of Science and Engineering, Final Report.* Task force on Women, Minorities, Handicapped in Science and Technology, January 1990.

43. Resnick, Lauren B. *Education and Learning to Think.* Committee on Mathematics, Science and Technology Education, National Research Council. Washington, DC: National Academy Press, 1987.

44. Schoenfeld, Alan H. (Ed.). *A Source Book for College Mathematics Teaching.* Washington, DC: Mathematical Association of America, 1990.

45. Senechal, Lester (Ed.). *Models for Undergraduate Research in Mathematics.* MAA Notes No. 18. Washington, DC: Mathematical Association of America, 1990.

46. Seymour, Elaine; Hewitt, Nancy M. *Factors Contributing to High Attrition Rates Among Science, Mathematics and Engineering Majors.* Bureau of Sociological Research. Boulder, CO. Westview Press, 1994, 1997.

47. Smith, David A.; Porter, Gerald J.; Leinbach, L. Carl; and Wenger, Ronald H. (Eds.) *Computers and Mathematics: The Use of Computers in Undergraduate Instruction.* MAA Notes No. 9. Washington, DC: Mathematical Association of America, 1988.

48. Steen, Lynn Arthur. Quantitative Literacy: Why Numeracy Matters for Schools and Colleges. *Focus* 22:2 (February 2002). Available at www.maa.org/features/QL.html.

49. Steen, Lynn Arthur. Twenty Questions that Deans Should Ask Their Mathematics Departments. *Bulletin of the American Association of Higher Education* 44:9 (May 1992).

50. Steen, Lynn Arthur (Ed.). *Calculus for A New Century: A Pump, Not a Filter.* MAA Notes No. 8. Washington, DC: Mathematical Association of America, 1988.

51. Steen, Lynn Arthur (Ed.). *Challenges for College Mathematics: An Agenda for the Next Decade.* Washington, DC: Focus, 10:6 (November-December 1990).

52. Steen, Lynn Arthur (Ed.). *Heeding the Call for Change: Suggestions for Curricular Action,* MAA Notes No. 22, Washington, DC: Mathematical Association of America, 1992.

53. Steen, Lynn Arthur (Ed.). *Library Recommendations for Undergraduate Mathematics.* MAA Reports Number 4, Washington, DC: Mathematical Association of America, 1992.

54. Steen, Lynn Arthur (Ed.). *Two-Year College Mathematics Library Recommendations.* MAA Reports Number 5, Washington, DC: Mathematical Association of America, 1992.

55. Sterrett, Andrew (Ed.). *Using Writing to Teach Mathematics.* MAA Notes No. 16. Washington, DC: Mathematical Association of America, 1990.

56. The Mathematical Association of Two-Year Colleges. *Crossroads in Mathematics: Standards for Introductory College Mathematics Before Calculus.* Memphis, TN: The American Mathematical Association of Two-Year Colleges, 1995. Available at www.imacc.org/standards/.

57. The American Mathematical Association of Two-Year Colleges. *Guidelines for Mathematics Departments at Two-Year Colleges.* Memphis, TN: American Mathematical Association of Two-Year Colleges, 1993.

58. Thurston, William P. Mathematical Education. *Notices of the American Mathematical Society,* 37:7 (September 1990) 844–850.

59. Treisman, Philip Uri. "A Study of the Mathematics Performance of Black Students at the University of California, Berkeley." In *Mathematicians and Education Reform.* Providence, RI: CBMS issues in Mathematics Education, Volume 1, 1990, pp. 33–46.

60. Tucker, Alan. *Recommendations for a General Mathematical Sciences Program.* Washington, DC: Mathematical Association of America, 1981. (Reprinted as first of six chapters of *Reshaping College Mathematics,* MAA Notes No. 13, 1989.)

61. Tucker, Alan (Ed.). *Models That Work: Case Studies in Effective Undergraduate Mathematics Programs.* MAA Notes No. 38. Washington, DC: Mathematical Association of America, 1996.

62. Tucker, Thomas (Ed.). *Priming the Calculus Pump: Innovations and Resources.* MAA Notes No. 17. Washington, DC: Mathematical Association of America, 1990.

63. Velez, William Yslas. Academic Advising as an Aggressive Activity. *Focus,* 14:4 (August 1994) 10–12.

64. Velez, William Yslas. Integration of Research and Education. *Notices of the American Mathematical Society,* 43:10 (October 1996) 1142–1146.
65. Wilf, Herbert S. Self-esteem in Mathematicians. *The College Mathematics Journal,* 21:4 (September 1990) 274–277.
66. Zorn, Paul. Computing in Undergraduate Mathematics. *Notices of the American Mathematical Society,* 34 (October 1987) 917–923.

Appendix A

Quantitative Reasoning for College Graduates: A Complement to the Standards

SUMMARY

What quantitative literacy requirements should be established for all students who receive a bachelor's degree? Over the years, the Mathematical Association of America (MAA) has approached this question in various ways, most recently by establishing, in 1989, a Subcommittee on Quantitative Literacy Requirements (henceforth called the Subcommittee) of its Committee on the Undergraduate Program in Mathematics. The work of the Subcommittee has been similar in some respects to the efforts of the National Council of Teachers of Mathematics (NCTM) that led to its celebrated Curriculum and Evaluation Standards for School Mathematics (1989) and related publications. The recommendations from the Subcommittee can be considered to complement those in the Standards. They also should be viewed as a reasonable extension of a Standards-based high school experience to the undergraduate level.

The Subcommittee began with the perception, supported by many recent studies and reports, that general mathematical knowledge among the American people is in a sorry state. It assumed that colleges and universities would welcome some suggestions on what they might do about the situation.

The discussions and investigations conducted by the Subcommittee led to four primary conclusions. The conclusions embody a vision that goes well beyond present practice in most places.

Conclusion 1. Colleges and universities should treat quantitative literacy as a thoroughly legitimate and even necessary goal for baccalaureate graduates.

Many authoritative mathematical and other groups have affirmed the importance of quantitative, or mathematical, skills in the population at large. These skills are valuable in various ways (this report lists nine), e.g. in daily life, further education, careers, and overall citizenship. To some degree these skills are acquired by the end of secondary education, but the post-secondary experience should reinforce what has been learned in school and go beyond. Thus the Subcommittee's concern has been not with quantitative literacy in general, but with quantitative literacy for college graduates, which naturally should differ in both depth and quality from that expected of high school graduates.

Conclusion 2. Colleges and universities should expect every college graduate to be able to apply simple mathematical methods to the solution of real-world problems.

Rote and passive learning of mathematical facts and procedures is not enough. Educated adults should be able to interpret mathematical models, represent mathematical information in several ways, and use different mathematical and statistical methods to solve problems, while recognizing that these methods have limits. These elements extend those in the ideal of "mathematical power" presented in the NCTM Standards, which include "methods of investigating and reasoning, means of communication, and notions of context." At the same time, these goals seem attainable.

Conclusion 3. Colleges and universities should devise and establish quantitative literacy programs each consisting of foundation experience and a continuation experience, and mathematics departments

should provide leadership in the development of such programs.

A required course or two is not sufficient. A student becomes quantitatively literate through a broad program that instills certain "long-term patterns of interaction and engagement." The program, the central idea of these recommendations, starts with a "foundation experience" into which students are appropriately placed and in which a carefully chosen course or two can raise entering students to a level of proficiency where they can benefit from the next phase, which is the "continuation experience."

In the continuation phase, later in their undergraduate programs students exercise and expand the elements of quantitative literacy they have already learned in the foundation experience and elsewhere. This phase is made possible by a framework of mathematics across the curriculum, an array of courses (both within and outside mathematics) and other educational experiences designed, in content and style, to contribute to the strengthening of quantitative literacy. The mathematics should be taught in context. Instructional materials should be current, practical, and conducive to active student involvement. Writing, student collaboration, and thoughtful use of instructional technology all have potentially important places. The program may also include the provision of mathematics clinics and other such resources.

In the course of these efforts, the needs, backgrounds, and expectations of people who in the past have tended to have special problems with mathematics should not be overlooked. Indeed, a well-designed quantitative literacy program may be of exceptional benefit to those persons who have special difficulties with mathematics.

Conclusion 4. Colleges and Universities should accept responsibility for overseeing their quantitative literacy programs through regular assessments.

A quantitative literacy program should be managed watchfully. At appropriate times and in appropriate ways, the results should be evaluated so as to obtain enlightened, realistic guidance for improvement. Evaluation methods should reflect course goals and teaching methods used, and besides pointing to possible improvements in the program can themselves be educationally beneficial. In particular, the evaluation methods should involve clearly applicationsoriented tasks.

[The report concludes with five appendices including references, a list of topics on which one might base a reasonable syllabus, brief descriptions of some existing foundations courses, a questionnaire for assessing attitudes toward mathematics, a list of problems related to minimal competency, a set of project ideas, several scoring guides, and comments on approaches to quantitative literacy for two specific majors.]

Committee on the Undergraduate Program in Mathematics. "Quantitative Reasoning for College Graduates: A Complement to the Standards". A Report of CUPM. Washington, D.C.: Mathematical Association of America, 1996.

Appendix B

The Undergraduate Major in the Mathematical Sciences

Summary

This CUPM report on the undergraduate major in the mathematical sciences describes a curricular structure with fixed components within which is considerable latitude in specific course choices. Combined with specialized curriculum concentrations or tracks within the major, this structure provides flexibility and utility. The structure involves both specific courses (e.g., "linear algebra"), and more general experiences (e.g., "sequential learning") derived through those courses. By making appropriate choices within components, students can obtain a strong major for prospective secondary teaching or for graduate school preparation.

The component structure with tracks is typical of the pattern of many of today's undergraduate mathematical sciences departments in that it allows many curricular choices. Seven components form the structure of the mathematical sciences major:

A. Calculus (with differential equations)
B. Linear algebra
C. Probability and statistics
D. Proof-based courses
E. An in-depth experience in mathematics
F. Applications and connections
G. Track courses, departmental requirements and electives

In addition to courses and components, the mathematical sciences major should also involve a variety of other types of experiences and activities that are, in some cases, "co-curricular." Several supportive activities are specifically cited as contributing to students' self-confidence and ability to work with others: integrative experiences, communication and team learning, independent mathematical learning and structured activities. The statements of philosophy in the report embody educational principles that can lead to an enriching educational experience and the recommended program structure provides a flexible vehicle for fulfilling those principles. One underlying tenet, however, transcends the particular form of curriculum implementation: It is only by requiring substantive achievement of our students that we will be able to produce the sort of quantitatively expert individuals who are going to be the mainstay of the discipline and of society for the next century."

Committee on the Undergraduate Program in Mathematics. "The Undergraduate Major in the Mathematical Sciences". A Report of CUPM. Washington, DC: Mathematical Association of America, 1991.

Appendix C

A Call for Change: Recommendations for the Mathematical Preparation of Teachers of Mathematics

Summary

"A Call for Change" is a set of recommendations for the mathematical preparation of teachers from the Mathematical Association of America. The document, recognizing that there are complex interactions among the teacher, the mathematics content being taught, and the students, speaks to recommended changes in the teaching and learning of mathematics by teachers. In this sense, it should be considered together with NCTM's document, "Professional Standards for Teaching Mathematics." This latter document addresses means to close the gap between recommended ideals of teaching mathematics and the reality of mathematics education in the schools today.

"A Call for Change" has four main sections:

Standards common to the preparation of mathematics teachers at all levels.

Standards in this section encompass the preparation recommended for mathematics teachers in order that they:
- principles;
- communicate mathematics accurately, both verbally and in writing;
- modeling;
- understand and use calculators and computers appropriately in the teaching and learning of mathematics;
- appreciate the development of mathematics both historically and culturally.

Standards for Teachers at the Elementary Level (K–4)

A core of experiences described with four broad standards on
- nature and use of number;
- geometry and measurement;
- patterns and functions;
- collecting, representing, and interpreting data.

Standards for Teachers at the Middle Grades Level (5–8)

At this level, the core experiences are described through five standards:
- number concepts and relationships;
- geometry and measurement;
- algebra and algebraic structures;
- probability and statistics;
- concepts of calculus.

Standards for Teachers at the Secondary Level (9–12)

The equivalent of a major in mathematics, but one quite different from that currently in place in most institutions, is recommended at this level. It is expected that the courses offered by departments of mathematics include the experience necessary to meet the common standards listed above. The Standards given in "A Call for Change" describe broad knowledge and understanding of mathematics. It is not the intention that a given standard describes the content of a single college-level mathematics course.

Why is there a need for change?

Mathematics continues to be a dynamic, changing discipline. There is new mathematics that can be exciting for young people to learn and technology provides new approaches for teachers to engage students in the teaching and learning of mathematics. The mathematics preparation of teachers must adapt to these changing realities.

Committee on the Mathematics Education of Teachers. "A Call for Change: Recommendations For The Mathematical Preparation Of Teachers Of Mathematics". James R. C. Leitzel, Editor. MAA Notes and Reports Series. Washington, D.C.: Mathematical Association of America, 1991.

Appendix D

Providing Resources for Computing in Undergraduate Mathematics

Computers and calculators are transforming the world in which students will live and work. Moreover, technology is changing the way mathematicians work and teach, as well as enhancing the potential for learning mathematics. These changes present both opportunities and challenges for college and university departments of mathematics.

The Mathematical Association of America urges colleges and universities to respond aggressively to the changing needs of their students. In particular, all mathematics departments should prepare students to use mathematics in a technological environment. To achieve this objective, faculties, departments, and institutions must work together:

- To ensure that all students have sufficient access to computing resources appropriate to the needs of their mathematics courses.
- To provide mathematics faculty members effective access to appropriate computing equipment.
- To provide faculty members with adequate time, opportunity, and professional incentives to use technology effectively.
- To provide the resources required for a computer enriched teaching environment.
- To provide effective technical support to departments of mathematics.

Faculties and administrations must together devise appropriate local solutions to the many problems that arise as mathematics departments adapt to the new role of calculators and computers. These problems include hardware (cost, access, location, ownership), software (effectiveness, licensing, hardware environments), space (laboratories, classrooms, offices); personnel (installation, maintenance, consulting); management (central vs. distributed); and work loads (course development, laboratory instruction).

Further information on development of effective calculator and computer environments for undergraduate mathematics can be obtained from The Mathematical Association of America, 1529 Eighteenth Street, NW, Washington, DC 20036.

Statement by the Board of Governors of the Mathematical Association of America, January 15, 1991, San Francisco, California.

Appendix E

Advising

Unlike an earlier, simpler day when all mathematics majors took the same sequence of courses with only a few electives in the senior year, the typical undergraduate mathematical sciences department today requires students to make substantial curricular choices. As a result, departments have advising responsibilities of a new order of magnitude. Students need departmental advice as soon as they show interest in (or potential for) a mathematics major. Advisors should carefully monitor each advisee's academic progress and changing goals, and together they should explore the many intellectual and career options available to mathematics majors. Career information is important. If a "minor" in another discipline is a degree requirement or option, then achieving the best choice of courses for a student may necessitate coordination between the major advisor and faculty in another department.

Advisors should pay particular attention to the need to retain capable undergraduates in the mathematical sciences pipeline, with special emphasis on the needs of underrepresented groups. When a department offers a choice of several mathematical tracks within the major, advisors have the added responsibility of providing students with complete information even when students do not ask many questions. Track systems may lead students to make lifetime choices with only minimal knowledge of the ramifications; therefore, departments utilizing these systems for their majors must assure careful and timely information. One requisite of an individualized approach to advising is that each advisor be assigned a reasonable number of advisees.

Committee on the Undergraduate Program in Mathematics. "The Undergraduate Major in the Mathematical Sciences ," Page 5. A Report of CUPM. Washington, D.C.: Mathematical Association of America, 1991.

Appendix F

Task Force to Review the 1993 MAA Guidelines for Programs and Departments in Undergraduate Mathematical Sciences

Donald L. Bentley, Pomona College, Claremont, CA

Sylvia T. Bozeman, Spelman College, Atlanta, GA

Chair: **John D. Fulton**, Virginia Polytechnic Institute and State University, Blacksburg, VA

Wanda L. Garner, Chair, Division of Math, Science and Engineering, Cabrillo College, Aptos, CA

Nancy L. Hagelgans, Ursinus College, Collegeville, PA

David J. Lutzer, College of William and Mary, Williamsburg, VA

Dale H. Mugler, University of Akron, Akron, OH

Barbara L. Osofsky, Rutgers University, New Brunswick, NJ

Jimmy L. Solomon, Dean, Paulson College of Science and Technology, Georgia Southern University, Statesboro, GA

Ann E. Watkins, California State University, Northridge, Northridge, CA

About the Editors

Tina Straley received a PhD degree in mathematics, specializing in combinatorics and universal algebra from Auburn University. She has been on the faculties of Kennesaw State University, Auburn University, and Spelman College. At Kennesaw she served as Chair of the Department of Mathematics and as Associate Vice President for Scholarship and Dean of Graduate Studies. Tina spent a year as Visiting Research Associate at Emory University and two years as Program Officer in Undergraduate Education at the National Science Foundation.

She has authored papers in combinatorics and graph theory and has had many grants from public and private organizations, including an NSF Trusteeship for her graduate education and support for undergraduate education and faculty development.

Dr. Straley has been active in several professional societies, especially the Mathematical Association of America, in which she was Southeastern Section Chair and Editor of the MAA Notes Series. She assumed her present position as Executive Director of the MAA in January 2000.

Marcia P. Sward served as MAA Executive Director from 1989 through 1999. She received her PhD from the University of Illinois, Champaign-Urbana, with a thesis in partial differential equations. She then joined the faculty of Trinity College and taught in the mathematics department for eleven years. In 1980, she was appointed as the MAA's first Associate Director and later also as the Administrative Officer of the Conference Board of the Mathematical Sciences. Dr. Sward left the MAA for four years to launch the Mathematical Sciences Education Board at the National Research Council, then returned to become the MAA's Executive Director. After retiring from the MAA, she served as Senior Director of Environmental Education at the National Environmental Education Foundation and is now Director of Environmental Education at the Audubon Naturalist Society.

Jon Scott is Professor of Mathematics at the Germantown, Maryland campus of Montgomery College. He was previously at the Takoma Park campus, where he served two terms as department chair. Jon recently won a College award for outstanding teaching and service. An active MAA member with a long-time interest in professional development, he is a member of the team that conceived and manages the MAA PREP program. Jon was Visiting Mathematician at the MAA offices during a 1995–96 sabbatical year. He co-directs, with Tina Straley and Arnie Ostebee, the PREP workshops for chairs of mathematics sciences departments.